Climate-Just Behavior

T0174741

This book highlights the obstacles to and potential for a just transformation as a way out of the current climate crisis.

This volume examines the barriers, opportunities and incentives around the pursuit of climate-just behavior, based on a comprehensive interdisciplinary and integrative analysis. It investigates how the gap between expressing concern about the climate crisis and giving it a high priority within the context of everyday behavior can be overcome. At the same time, it looks at the challenging politico-economic framework conditions such as the strong economic growth and profit orientation of capitalism. Although justice is a fundamental human motive, which should induce climate-just behavior, system justification is common and makes people rather justify their unjust behavior. In this book, a general and systemic framework on human behavior is provided, including internal factors, such as knowledge and psychological needs, external factors, such as socio-cultural and politico-economic factors, feedback loops and interactions. The authors draw on multiple theories to examine how denial and moral disengagement affect individual responsibility, despite real-world evidence of the climate crisis. The book highlights the role of emotions in encouraging a pro-environmental response and discusses solutions on both the individual and the collective level, such as transparency laws. Moreover, making climate-friendly options more accessible, affordable and convenient facilitates behavior change more effectively. Overall, this book presents knowledge-based, realistic approaches to surmounting these obstacles in order to achieve a more climate-just world.

Climate-Just Behavior will be of great interest to students and scholars of climate change, climate justice, environmental geography and environmental psychology.

Susanne Stoll-Kleemann is a University Professor and currently works as Chair of Sustainability Science and Applied Geography, University of Greifswald, Germany. There, she heads the Master's program "Sustainability Geography". She previously conducted research on the psychology of climate-friendly behavior at the Potsdam Institute for Climate Impact Research (Germany) and at the Swiss Federal Institute of Technology in Zurich, Switzerland.

Susanne Nicolai is a psychologist and currently pursuing her doctoral studies on the perception of injustice and moral emotions amidst the climate crisis at the University of Greifswald. Additionally, she actively engages in social movements advocating for climate justice.

Routledge Focus on Environment and Sustainability

For more information about this series, please visit: www.routledge.com/Routledge-Focus-on-Environment-and-Sustainability/book-series/RFES

Climate-Just Behavior

Foundations and Transformational
Approaches

**Susanne Stoll-Kleemann
and Susanne Nicolai**

Routledge
Taylor & Francis Group
LONDON AND NEW YORK

from Routledge

First published 2024
by Routledge
4 Park Square, Milton Park, Abingdon, Oxon OX14 4RN

and by Routledge
605 Third Avenue, New York, NY 10158

Routledge is an imprint of the Taylor & Francis Group, an informa business

© 2024 Susanne Stoll-Kleemann and Susanne Nicolai

British Library Cataloguing-in-Publication Data
A catalogue record for this book is available from the British Library

ISBN: 978-0-367-47116-3 (hbk)
ISBN: 978-1-032-84024-6 (pbk)
ISBN: 978-1-003-03354-7 (ebk)

DOI: 10.4324/9781003033547

Typeset in Times New Roman
by KnowledgeWorks Global Ltd.

Susanne Stoll-Kleemann dedicates her work to all those people who already (sometimes unfortunately involuntarily) have a low-carbon footprint or are on their way to soon enjoying the pleasure of not living at the expense of present and future generations.

Susanne Nicolai wants to dedicate her work to her wonderful husband. Thank you for your unwavering support and presence by my side.

Both authors extend their gratitude to Laura Hoffmann, whose assistance as a student helper greatly contributed to the layout and finalization of this project. We are grateful to our colleagues Amelie Michalke, Lennart Stein, Alexandra Kruse, Wanda Born, Anne-Cristina de la Vega-Leinert and Wibke Müller for their time and insightful contributions to several chapters. Additionally, they express appreciation to Jyotsna Gurung and Annabelle Harris for their support throughout this endeavor.

Disclaimer: The authors of this book are both white individuals residing in an industrialized country (Germany). While they substantiate their assertions with scientific evidence, they acknowledge the potential bias inherent in their privileged worldview, which may have influenced their literature research or conclusions. Nonetheless, both authors have diligently crafted the manuscript in accordance with T&F's editorial principles of diversity, equity and inclusion and remain committed to advancing social justice.

Contents

1 Overview and Introduction

Susanne Nicolai

Relevance and Definition of Climate Justice

Climate change encompasses global warming as well as its impacts on the Earth's climate system. Human activities such as burning fossil fuels, deforestation, and certain agricultural and industrial practices release greenhouse gasses like carbon dioxide and methane, intensify the natural greenhouse effect and cause global warming.

The consequences of climate change are evident in expanding deserts and more frequent heat waves, wildfires and extreme weather events. Melting permafrost, glacial retreat and rising sea levels are contributing to environmental changes. Ecosystems are rapidly changing, which leads to species relocation and threats of extinction. Nonetheless, human land use possibly plays a greater role in species extinction.

Climate change poses severe threats to human populations, leading to flooding, heatwaves, food and water scarcity, diseases and economic losses. The World Health Organization considers it the greatest threat to global health. While adaptation measures like flood control and drought-resistant crops can mitigate some risks, vulnerable countries face significant challenges given that adaptation requires resources, technology and power.

Ongoing climate change threatens all three dimensions of human rights: (1) civil and political rights; (2) economic, social, and cultural rights; and (3) collective rights (Schapper, 2018). Furthermore, climate change is causing major impacts on human lives, health and livelihoods, particularly among the most marginal populations of the world. Droughts and floods undermine food security, with extended hunger and poverty as consequences. The Intergovernmental Panel on Climate Change (IPCC) reports with medium to high confidence that climate change is already driving negative impacts on health and well-being, for example, through malnutrition and heat (IPCC (AR6), 2022). Researchers estimate that there were 114 heat-related deaths per million inhabitants in Europe during the summer of 2022 (Ballester et al., 2023). Higher temperatures increase the risk of heart attacks, which are already the most prominent cause of death worldwide. Moreover, heat also increases

DOI: 10.4324/9781003033547-1

other health risks such as respiratory diseases, kidney failure and cognitive impairment. The 2003 heat wave claimed up to 70,000 lives in Europe, and climate change will make such events much more common in the future, thus contributing to more heat-related deaths. According to the Lancet Countdown (Watts et al., 2018) – an association of international scientists – there could be 30,000 heat-related deaths per year in the EU by 2030, and by 2080 as many as 110,000 people a year could die from heat. Droughts exacerbated by anthropogenic climate change have already instigated conflicts over depleting water availability and eroding arable land.[1]

The persistently high carbon footprint of individuals, collectives, economic sectors and national states leads to a worsening climate crisis and thus climate injustice in the form of deaths from extreme weather events such as droughts, floods, malnutrition, heat waves and climate wars, as well as migration and increasingly unmanageable and uninhabitable regions on Earth.

The Paris Agreement aims to limit warming to *well below* 2°C, although current estimates still point to a temperature rise of around 2.7°C by the end of the century. Achieving the 1.5°C target implies halving emissions by 2030 and reaching net zero emissions by 2050. Reducing greenhouse gas emissions is also called mitigation, while preparing infrastructure for a warmed world is termed as adaptation. The IPCC (2022) emphasizes the need for integrated responses that link mitigation and adaptation, as many options can help to address climate change but no single option is sufficient by itself. The effective implementation of both adaptation and mitigation strategies depends on policies and cooperation at all scales (IPCC, 2022). However, in this book, we focus on the driver of climate change – greenhouse gas emissions – and therefore mitigation strategies. Greenhouse gasses need to be reduced rapidly, effectively and in a socially just manner. From our perspective, *climate-just behavior* is critical to achieve the systemic transformation required to address key drivers of anthropogenic climate change and includes mitigation strategies that can be performed at the individual level. While other authors use terms such as *climate-friendly behavior* or *low-carbon behavior*, we use the term *climate-just behavior* in this book as we want to emphasize that we see this kind of behavior as the transformation to climate justice.

Williamson et al (2018, p. 14) define 30 possible behaviors to mitigate climate change, although different behaviors strongly vary in their effectiveness. Table 1.1 displays the top ten most impactful behaviors from their list.

Williamson et al. (2018) estimated that by applying all 30 solutions to reduce emissions from human consumption, individuals would be able to save 393–729 $GtCO_2$-equivalents or 19.9% (plausible) to 36.8% (optimum) of emissions mitigated. As the authors note, "it is important to highlight that many of these solutions are beneficial not only in terms of their mitigation potential but also in terms of economics, human health, and well-being" (Williamson et al., 2018, p. 18). However, we are well aware that some individuals

Table 1.1 Top 10 of the 30 solutions to reduce emissions from human consumption across major economic sectors and solutions adoption scenarios

Sector	Solution	Description	Plausible-optimum Scenario emissions Reduction $(GtCO_2\text{-}eq)$
Food	1 Reduced food waste	Minimizing food loss and wastage throughout the food supply chain from harvest to consumption	70.5–93.7
	2 Plant-rich diets	Eating more plant-based foods and fewer animal-based proteins and products (e.g. meat, dairy)	66.1–87.0
Agriculture and land management	3 Silvopasture	Adding trees to pastures to increase productivity	31.2–65.0
	5 Tropical staple trees	Growing trees and other perennial crops for staple proteins, fats and starches	20.2–47.2
	7 Tree intercropping	Growing trees together with annual crops in a given area at the same time	17.2–37.0
	8 Regenerative agriculture	Adopting at least four of the following six agricultural practices: compost application, cover crops, crop rotation, green manures, no-till or reduced tillage and/or organic production	23.2–32.4
	9 Farmland restoration	Restoring degraded, abandoned farmland to grow crops or native vegetation	14.1–30.8
	10 Managed grazing	Adjusting stocking rates, timing and intensity of grazing in grassland soils	16.3–27.9
Transportation	4 Electric vehicles	Driving battery and plug-in vehicles instead of conventional vehicles	10.8–52.4
Energy and materials	6 Rooftop solar	Installing rooftop photovoltaic systems under 1 MW	24.6–40.3

Source: Own representation based on Williamson et al. (2018).

are able to reduce a larger amount of emissions as their lifestyle emits considerably more greenhouse gasses compared with others.

In this book, we follow Parfit's (2011) distinction between *subsistence emissions* and *luxury emissions* (Shue, 1993, 2001). Peeters et al. (2019) specify that "while individuals cannot be faulted for the greenhouse gases (GHG) they emit in order to meet their basic needs, and many of their choices are determined by contextual factors that are beyond their control (such as infrastructure, and the energy regime) [...] they can be held accountable for their profligate emissions" (p. 427). This leads to a central theme of this book, namely that individual emitters are responsible for the climate crisis because they can decide to reduce *unnecessary* greenhouse gas emissions, for example by reducing their consumption of animal products, flying less or "reducing their energy consumption by undertaking fairly low-cost and easy actions for which feasible alternatives exist – emissions that would unambiguously classify as luxury emissions" (Peeters et al., 2019, p. 427).

Therefore, in recent years it has become increasingly clear that the climate crisis is also – or above all – a question of justice. Additionally, the average carbon footprint of someone among the richest 1% of the world's population is estimated to be 175 times that of someone in the poorest 10% (Otto et al., 2019). For example, according to current estimations, the global wealthiest people only represent 0.54% of the world's population but are responsible for 13.6% (3.9 billion tons of carbon emissions per year) of the annual global carbon emissions, whereas the poorest 50% of the world are only responsible for 10% of the annual global carbon emissions (Otto et al., 2019).

The term climate justice incorporates various principles of the concept of justice (Chapter 2) that apply to the domain of climate change. Moreover, the term is strongly context- and group-dependent. For example, the term is used differently by social movements, NGOs, academics, and policy-makers (Schlosberg & Collins, 2014). This is due to the fact that the term has been constructed and emerged through many players, and above all social movements. As Schlosberg and Collins (2014) highlight, the history of the terms climate justice and environmental justice (a precursor to climate justice) dates back to the Warren County events in the United States in 1982, where the poor, mainly African-American community mobilized against the dumping of high-toxic waste. Although environmental concerns have been part of social movements of various minorities for even longer, "the environmental justice movement was more than simply a merger between civil rights and environmental groups – it included the occupational health and safety movement, the Indigenous land rights movement, the public health and safety movements, and various social and economic justice movements" as well as urban environmental groups (Schlosberg & Collins, 2014, p. 360). This term, therefore, carries significant power because it has the potential to unify social and environmental movements: activists who occupy seemingly disparate issues such as peace, human rights or the environment can unite behind the demand for climate justice.

However, a workable conceptualization is needed to make the term climate justice clear and understandable. Meikle et al. (2016) from Glasgow's Caledonian University Centre for Climate Justice compared various definitions of climate justice that are used by different groups. In their review, they describe a range of approaches involved in climate justice, including the recognition of triple inequality: in responsibility, vulnerability and mitigation. The authors argue that a commitment to reparations and fair distribution of the world's wealth is necessary to dismantle the fossil fuel corporate power structure and realize a vision that dissolves and alleviates the unequal burdens created by climate change. In an attempt to find a workable conceptualization of climate justice, Meikle et al. (2016) recommend Jafry's (2018) definition: "Climate justice recognizes humanity's responsibility for the impacts of greenhouse gas emissions on the poorest and most vulnerable people in society by critically addressing inequality and promoting transformative approaches to address the root causes of climate change" (Jafry, 2018, p. 3).

This stresses that poverty, inequality, gender inequality and power imbalances act as *multipliers* for the negative impacts of climate change.

Consequently, Klinsky et al. (2012) state that it is highly difficult to agree on a just division, as "countries vary enormously by opportunities to reduce emission, past emission history, projected emission trajectories and financial resources, and good arguments about justice can be made on all of these dimensions" (p. 862).

However, climate justice is not only about international relations and dependencies or social groups, as it also considers intergenerational injustices. As the benefits and burdens for today's youth and future generations are mostly based on models and predictions, answering questions regarding a fair distribution (or "distributive justice", Chapter 2) is even more difficult. Nonetheless, Roser and Seidel (2017) schematically show the effects of today's emissions on future generations via a higher atmospheric concentration and in turn a temperature increase (Figure 1.1). Later in this chapter, we show why youth (as well as the elderly) are categorized as a specific vulnerable group.

Another important aspect of climate injustice is the historical perspective: the cumulative amount of GHG that have been emitted to date varies across different states. It is often argued that China or India are high emitters at present, although for a long time, EU countries, the United States and Canada were the prime emitters of greenhouse gas emissions. Therefore,

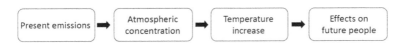

Figure 1.1 Schematic representation of the chain of effects of today's greenhouse gas emissions on future generations (own representation based on Roser & Seidel, 2017)

it is useful to look at cumulative greenhouse gas emissions. The concept of historic debt in greenhouse gas emissions – also known as climate debt – refers to the idea that certain countries or groups of people have a responsibility to compensate for the negative impacts of climate change caused by their past or current greenhouse gas emissions. This concept is based on the idea of historical accountability and can be estimated based on actual and projected emissions and the social cost of carbon. Historic debt in climate emissions is a concept that has been widely debated in the literature. Blomfield (2019) rejects the notion of historical emissions debt, arguing that there is no fair shares principle for the historical use of the climate sink. Godard (2012) similarly criticizes the concept, highlighting its ill-founded nature and its potential to hinder forward-looking agreements. By contrast, Duus-Otterström (2013) proposes the Inherited Debt Principle as a way to address the problem of past emissions, which considers the role of historical injustice in shaping current duties. Blomfield (2019) criticized it for being too simplistic and focusing on blaming instead of finding solutions. The journalistic collective carbon brief published a highly impressive analysis on these cumulative emissions (from 1850 to 2021; Evans, 2021), which can be found in Figure 1.2. Evans further argues that a historical perspective matters because the cumulative amount of carbon emissions since the start of the Industrial Revolution is closely tied to the 1.2°C of warming that has already occurred. Using this approach, the United States remains the most prominent emitter of GHG by far.

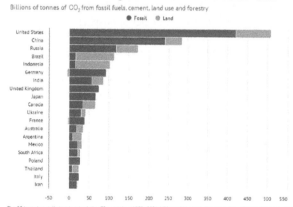

The countries with the largest cumulative emissions 1850–2021

Billions of tonnes of CO_2 from fossil fuels, cement, land use and forestry

● Fossil   ● Land

The 20 largest contributors to cumulative CO_2 emissions 1850–2021, billions of tonnes, broken down into subtotals from fossil fuels and cement (grey) as well as land use and forestry (green). Source: Carbon Brief analysis of figures from the Global Carbon Project (https://www.globalcarbonproject.org/), CDIAC (https://cdiac.ess-dive.lbl.gov/), Our World in Data (https://ourworldindata.org/), Carbon Monitor (https://carbonmonitor.org/), Houghton and Nassikas (https://agupubs.onlinelibrary.wiley.com/doi/abs/10.1002/2016GB005546) (2017) and Hansis et al at (https://agupubs.onlinelibrary.wiley.com/doi/10.1002/2014GB004997) (2015). Chart by Carbon Brief using Highcharts (https://www.highcharts.com/).

Figure 1.2 Countries with the largest cumulative emissions from 1850–2021 (calculated by Carbon Brief).

The Role of the Individual in Relation to Politics and Society

Due to persistent delays in enforcing adequate climate protection regulations at the national and global scale and devising effective incentives, individuals, communities, companies, and civil society at large have a vital role to play in overcoming the injustices and harmful consequences of climate change. In this book, we focus on the individual level, namely high- and middle-income citizens, mainly in rich countries but also include global elites around the world. We examine personal transition paths toward reducing carbon-based emissions. However, system change at the societal level is necessary to make these effective. Structural injustices have led to the current situation and further exacerbate future climate impacts on the planet. Nevertheless, we contend that we need to understand human behavior in relation to climate mitigation at the individual level before considering behavior at the collective level. Nevertheless, climate mitigation needs to be pursued simultaneously at both the institutional and individual levels. As the IPCC (2022) stated, the window of opportunities to act is rapidly closing. Given that governments, corporations and societies are also made up of individuals, a promising approach to complement existing efforts toward climate mitigation is to focus on specific individuals who have disproportionate amount of power in society, such as politicians and particularly rich people. Our aim is to combine interdisciplinary approaches from Psychology, Sociology, Economics and some further social and behavioral sciences to identify opportunities toward individual and collective behavioral change, as a step toward a socio-ecological transformation.

Vulnerable Groups

The latest report by the IPCC (2022) suggests that it is not only natural conditions that determine who is most affected by the climate crisis but rather the living conditions of the individuals. Limited economic and basic resources (such as clean water), restricted political power, inequality, political instabilities and wars are only a few examples of structural factors that increase the vulnerability of individuals. Climate change disproportionately affects low-income countries, younger generations, women and people with low income more intensely than the high-income countries, elderly people, men, and people with high income (IPCC, 2022). These groups are (1) more vulnerable, (2) have fewer options to adapt to climate change consequences, (3) own less power and opportunities to decide the political agenda, and (4) are less responsible for the greenhouse gas emissions that lead to climate change (*climate debt*). It is important to note that vulnerability is not a trait, but rather it is mostly societal. If people with disabilities or other vulnerable groups are more at risk, it is not due to themselves but rather societal barriers, because protective measures do not work for them. However, individuals are not equal in their physiology, health and abilities, which is why society must find solutions to protect all of these groups.

Low-income Countries

Despite happening on a global scale, climate change does not affect all regions to the same extent. For example, the nonprofit organization *berkleyearth* shows with an online tool how much various countries have warmed until 2020.[2] Although the world has warmed 1.3°C until 2020 since 1850, the United States has already warmed 1.7°C and Slovakia even 2.3°C. Furthermore, climate change-induced extreme weather events such as heavy rain and following floods particularly affect specific regions like the Philippines, Pakistan, and Haiti (Eckstein et al., 2021). However, it is not only the geographical position that makes up higher vulnerabilities to climate change consequences, given that social factors also play a critical role in this regard.

To systematically assess what makes countries vulnerable to climate change, Brooks et al. (2005) analyzed 46 proxy indicators (factors) for vulnerability in the categories economy, health and nutrition, education, infrastructure, governance, geography and demography, agriculture, ecology and technology. With the help of weighted factors and a categorization of each country for each factor, they made out 11 key vulnerability factors that correlated with decadal mortality outcome at the 10% significance level. These key indicators are (1) population with access to sanitation, (2) literacy rate of the 15–24-year-olds, (3) maternal mortality, (4) literacy rate of the population over 15 years, (5) caloric intake, (6) voice and accountability, (7) civil liberties, (8) political rights, (9) government effectiveness, (10) literacy ratio (female to male) and (11) life expectancy at birth. Further, although gross national income (GNI) is not mentioned by Brooks et al., it stands in relation to these key indicators.

The 2021 Global Climate Risk Index (Eckstein et al., 2021) examines and categorizes the extent to which countries and regions have experienced the effects of climate-related extreme weather events such as storms, floods and heatwaves. The analysis takes into consideration the year 2019 as well as the period from 2000 to 2019. In 2019, Mozambique, Zimbabwe and the Bahamas were the most heavily impacted countries, while Puerto Rico, Myanmar and Haiti ranked the highest in the period spanning from 2000 to 2019. The Climate Risk Index Report stressed how, while climate change is intensifying globally, the poorest nations are disproportionately affected by the consequences of extreme weather events due to their heightened vulnerability, limited capacity to cope and longer recovery times. Further, the Global Climate Risk Index resumes that in total, during the period from 2000 to 2019, over 475,000 individuals lost their lives due to the direct impact of over 11,000 extreme weather events worldwide. Losses amounted to around US$2.56 trillion (in purchasing power parities; Eckstein et al., 2021). Eight out of the ten countries most affected by extreme weather events in 2019 belong to the low- to lower-/middle-income category (Eckstein et al., 2021).

In contrast to other authors, we do not use the term *developing countries*, as this is a very normative term that carries the risk of being interpreted as backwardness, underdevelopment or nondevelopment. This can be hurtful to the representatives of these countries. At the same time, it is assumed that these countries are really developing (in the sense of a Western- or Europe-centered worldview). To deal with this critique, newer terms of *global north* and *global south* are used by NGOs or in some academic literature. These terms are intended to describe the situation of countries in the globalized world as free of values and hierarchies as possible. In this sense, a country of the global south is a politically, economically or socially disadvantaged state. The countries of the global north, on the other hand, are in a privileged position in terms of prosperity, political freedom and economic development. However, these terms lead to confusion, as the terms north and south are used beyond the actual geographical location of countries. These new terms are also criticized for again taking the perspective of the industrialized countries while ignoring the BRICS[3] countries. We therefore use the term low-income countries, defined as those countries "which have the weakest economies when evaluated by the World Bank, [...] based upon their gross national income (GNI) per capita" (World Bank, 2023). These nations and regions are categorized into four groups ranging from low to high income, and the boundaries of these categories are annually adjusted to account for global inflation (World Bank, 2023). Of course, this classification also has problems, as it is only based on a single factor, namely income. However, from our perspective, this term is the most neutral one.

Nevertheless, the GNI does not account for wealth inequality and does not provide the differentiation needed to isolate elites. In our research context, elites are understood to be individuals who have a disproportionately large share of the available national economic resources, who, we contend, are better protected from the impacts of climate change whether in high or low-income countries. The Southern Center for Inequality Studies (SCIS) at the University of the Witwatersrand investigates wealth and income inequality in various regions of the world. For example, they analyzed that in South Africa wealth inequality did not improve but rather even worsened since the end of apartheid (1993). The SCIS further explains some mechanisms behind that inequality, for example the Brazilian tax system that exacerbates existing wealth inequality by operating in a highly regressive manner, thereby burdening the poor disproportionately compared to the wealthier segments of society. This system hampers social equality by impeding upward mobility and demonstrates a lack of attention to potentially redistributive taxes, such as those on property and financial gains (Bressan et al., 2023).

Low-income Individuals

Today, there is a worrying level of inequality both between and within countries. According to the World Inequality Report (Chancel et al., 2022), the deregulation and liberalization of markets during globalization have led to rapidly rising income and wealth inequality in almost all countries since the 1980s and shows how drastic this development is. While in 1965, the Chief Executive Officers (CEOs) of the largest listed companies earned 21 times the average salary of employees, in 2020 the ratio was 352:1. Beyond income, the greatest inequality relates to wealth: the poorer half of the world has only 2% of global wealth, while the richest 10% control 38% of the global wealth (Chancel et al., 2022).

On the one hand, if everyone in high-income nations emitted as little as the poorer 50% of the world's population, the climate targets set for 2030 would already have been met (Chancel et al., 2022). On the other hand, excessive consumption by the rich also exacerbates the climate crisis through social emulation effects. The greater the gap between rich and poor in a society, the more important social status becomes. In unequal societies, people from all income groups are more likely to fear for their status (Layte & Whelan, 2014) and worry about how others judge them. They want to and eventually buy more things (Kronenberg, 2023; Walasek & Brown, 2016).

Kronenberg (2023) also points to another link between inequality and the climate crisis: if inequality is excessive, societies are insufficiently resilient to achieve the necessary socio-ecological transformation, first because extreme inequality leads to the erosion of democracy. By concentrating wealth at the top of society, political power also shifts there. Thanks to their financial resources, a few can exert an excessive influence on politics and push through their interests. As a result, the democratic idea of fair representation is lost. Second, social tensions increase as inequality grows. When people feel disconnected and ignored, can no longer afford their rent or dare not turn up the heating on, mistrust and fears increase, causing the loss of social cohesion necessary for socio-ecological transformation.

Women

Women have different roles, (financial) resources, power (e.g. government mandates), rights, knowledge and time with which to cope with climate change, in comparison to men (e.g. Nelson, 2011). Due to these gender inequalities, they are disproportionately strongly affected and especially vulnerable by climate change (IPCC, 2022).

According to the Women's Environmental Network (Haigh & Vallely, 2010), women face escalating challenges related to climate change. They are not only more likely to live in poverty but also experience increasing hardships such as work and income losses, negative health impacts and violence following climate change-related disasters. Displacement and difficulties arise when other family members, usually male, migrate for economic reasons.

Roles, responsibilities and workloads of women and men vary globally based on wealth and social characteristics. However, it is widely observed that women generally have higher workloads due to domestic, reproductive and various economic responsibilities. Climate change exacerbates this workload, contributing to health problems among women, particularly in the face of events like droughts.

The effects of climate change, such as rising food prices and food shortages, disproportionately affect women, who are often the first to suffer. Gender roles and biological differences contribute to exacerbated health inequalities among women. Instances of violence, including sexual harassment, are prevalent during resource conflicts and following disaster events.

Furthermore, women are expected to adapt to the effects of climate change, increasing their already substantial workload. Unintended consequences of responses to climate change, such as forestry projects and biofuel production, further compound the challenges faced by women. In essence, women's vulnerability to climate change spans economic, health and violence-related dimensions, with the severity increasing across these interconnected challenges.

Haigh and Vallely (2010) also note while disadvantaged men are also vulnerable to climate change, the subordination of women is widespread, so that men are more privileged, even in poor communities. Nevertheless, as Nelson (2011) highlights, women should not only be seen as passive, vulnerable victims. Doing this would disregard their possible role as climate agents. For example, women across age classes and across 14 countries in Europe, Latin America and the United States show consistently stronger pro-environmental behavior, attitudes and concern than men (see review by Gifford & Nilsson, 2014).

Finally, men emit more carbon dioxide and are therefore more responsible for climate change, for example via driving longer distances by car (and owning a car), choosing more energy-intense vehicles, having a more emitting lifestyle (e.g. eating more meat), working in branches that emit more carbon emissions, and having higher incomes to spend on consumption (Pearse & Connell, 2015).

Younger Generations

Children and young people are affected in different ways. First, direct effects on both their physical health and mental health are likely to occur. For example, fatalities and injuries, heat-related illnesses, exposure to environmental toxins, infectious, gastrointestinal and parasitic diseases, which become more likely with higher temperatures, affect children and youth more strongly (Sanson et al., 2019). Furthermore, recent studies show that depression, anxiety, sleep problems, cognitive deficits and learning problems as well as posttraumatic stress disorders emerge with extreme weather events and the confrontation with climate change (Sanson et al., 2019). The term *climate anxiety* has been coined by Psychologists (e.g. Clayton & Karazsia, 2020; Wullenkord & Reese, 2021) to encompass such worries about climate change consequences.

For example, based on a study (Marks et al., 2021) of 10,000 children and adolescents from ten different countries around the globe, 75% agreed with the statement *The future is scar* and more than half (55%) thought they had fewer opportunities than their parents. Nearly half (45%) said they felt worried or anxious about the climate in a way that affected their daily lives and functioning. The negative view of the future also influences personal decisions: four out of ten people between the ages of 16 and 25 (39%) are undecided about whether to have children due to the climate crisis. The figure was particularly high in Brazil, where almost half (48%) of those surveyed are hesitant about having children. The study also shows negative evaluations of the climate policies of the governments of the participants and therefore relates climate anxiety to the inaction of political agents (Marks et al., 2021).

It should be noted that in some areas of the world (e.g. Brazil in the study above), climate change consequences are already extremely and regularly visible for some years. In other areas, children are consistently exposed to climate models and future predictions concerning extreme weather events and food shortages through education and media. However, they often witness a lack of corresponding actions from their parents and society. Given this contrast, it is reasonable that concerns about climate change are particularly prevalent among those who will inhabit the planet for several more decades, namely children and young people. It should be noted that young people in most affected areas also show the highest scores on concern in relation to young people in less affected areas (Marks et al., 2021).

Furthermore, more indirect effects of climate change disproportionately affect children, such as food shortages, intergroup conflict, economic dislocation and forced migration. Moreover, their development may be impaired due to destroyed schools or heightened domestic violence after extreme weather events (Sanson et al., 2019).[4] Therefore, climate change is increasing children's dependency on caring adults. United Nations Children's Fund (UNICEF) brings up the point that children already are the most affected group due to climate change, as, for example, 10 million children were displaced in 2020 alone due to extreme weather events. UNICEF further argues that children – regardless of where they live – need a clean and intact environment to develop well and healthily. The UN Convention on the Rights of the Child, adopted in 1989, does not yet include the right to a clean environment. However, it guarantees every child the right to life and adequate living conditions, whereby a healthy environment is essential for this (UNICEF, 2023).

Elderly

Nevertheless, on the opposite side, elderly people also belong to the most vulnerable groups due to for example physical impairment, mental condition, diminished sensory awareness, chronic health conditions as well as socioeconomic status. At the same time, age is not causally linked to vulnerability, elderly people rather need special preparations (Janmaimool & Watanabe, 2012). For example, elderly people suffer more under heat (Ballester et al.,

2023) and are often not adequately reached by risk communication on upcoming extreme weather events. It occurs that four factors determine whether elderly people are self-prepared to extreme weather events. Those are levels of education, the number of family members (negative predictor), the number of fire experiences and conflicts with family members (Janmaimool & Watanabe, 2012).

However, even climate mitigation programs affect young people the most. For example, to prevent global warming exceeding 1.5°C, net zero emissions by 2050 are needed. The International Energy Agency calculated children born in this decade will only be allowed to emit ten times less carbon than those born in the 1950s to achieve this (Cozzi et al., 2022). The restrictions on the lifestyle of this and future generations, therefore, will be immense if these restrictions are ever enforced.

Indigenous People

According to the United Nations (2024), Indigenous peoples are defined as those who have strong ties to their traditional habitats or ancestral territories. They identify themselves as part of a unique cultural group, descending from communities that inhabited these territories prior to the establishment of modern states and current borders. Typically, these groups preserve their cultural and social distinctiveness, as well as their social, economic, cultural and political institutions, often maintaining a degree of separation from the larger society or culture. According to World Bank statistics from 2019, it is estimated that approximately 370 million Indigenous peoples exist worldwide, constituting approximately 5% of the global population. Seventy percent of the global Indigenous population lives in Asia Pacific Territory, a region that is especially at risk to climate change impacts (Shaffril et al., 2020).

As argued by Wildcat (2013), Indigenous peoples and marginalized populations face heightened exposure and sensitivity to climate change impacts, primarily due to their resource-based livelihoods and the precarious locations of their homes. Living in small, rural communities marked by low socioeconomic conditions, they often experience political marginalization. Additionally, their substantial involvement in small-scale food production, such as agriculture and fisheries, directly ties them to the ecosystems of their land and territories (IPCC, 2022). Beyond economic ties, Indigenous peoples maintain profound spiritual and cultural connections with their waters and lands. Consequently, the repercussions of climate change extend beyond environmental impacts, affecting Indigenous people at a fundamental level of identity.

Nevertheless, Etchart (2017) contends that framing Indigenous peoples solely as victims of climate change is unjustified; instead, they should be recognized as active agents in environmental conservation. To illustrate, representatives from Indigenous communities across various regions have proactively sought opportunities to contribute to the fight against climate change. They have actively participated in international environmental conferences, engaged in activism and played roles in local and national politics to amplify their voices. This engagement is grounded in substantial reasons: While the territories of the world's 370 million Indigenous peoples cover 24% of the global land, they harbor 80% of the world's biodiversity. Indigenous communities view themselves as guardians of their land and water, successfully fulfilling this role.

Despite their pivotal role in environmental conservation, Indigenous peoples have historically been excluded from political climate decision-making processes. For instance, they were not permitted to contribute to the development of the Millennium Development Goals in 2000. However, through persistent efforts, Indigenous peoples have gradually gained inclusion in conferences and drafting processes, such as those shaping the Sustainable Development Goals. Notably, they actively confront and resist endeavors by oil companies and others seeking to harm their land. In summary, Etchart (2017) concludes that Indigenous peoples currently wield significant influence and are a force to be reckoned with.

Intersection

Furthermore, intersectional discrimination may manifest, involving the simultaneous discrimination of individuals based on multiple factors. For instance, the circumstances of children vary significantly depending on their families' financial resources. Moreover, those experiencing intersecting forms of discrimination often lack a platform where their voices are acknowledged; for instance, Indigenous women may not be heard by both white individuals and men within their own communities. The array of discriminatory factors is extensive, including considerations such as disability, psychological disorders, trauma, urban-rural disparities, as well as differences in knowledge and education.

Summary: What Is the Aim of Climate Justice?

Building on Gabbatiss and Tandon (2021), the aims of climate justice (and the climate justice movement) can be summarized in four points: fair share of emissions, no false solutions, just transition and a ban of fossil fuels and corporate interference in global climate contracts.

Fair Share of Emissions

Throughout history, fossil fuel usage and greenhouse gas emissions have been disproportionately concentrated in a small number of relatively affluent nations. The climate justice movement emphasizes the crucial need for wealthy countries to acknowledge this historical responsibility and take substantial action to limit global warming to 1.5°C.

This demand also considers the greater financial capacity that these nations possess to invest in decarbonization efforts. In this respect, the concept of *carbon budget* has emerged in climate science and policy discussions over the last decade to refer to the remaining level of emissions that may be released before a specific temperature threshold is passed. As of January 1, 2021, it has been estimated that there are approximately 460 billion tons of carbon dioxide ($GtCO_2$) left in the global carbon budget, the equivalent of approximately 11.5 years of emissions at 2020 levels, before reaching the level that implies a commitment to a warming level of 1.5°C.

No False Solutions

Technologies and policies that reduce emissions may nevertheless be false solutions if they do not support the broader goal of climate justice.

This position has been articulated in the 27 Bali Principles of Climate Justice, defined by various NGOs in 2002 (see corpwatch.org), contending that market-based mechanisms and technological *fixes* promoted by transnational corporations are misleading solutions that worsen the problem instead of solving it. Geoengineering[5] or nuclear power are examples for such unjust (and unsafe), therefore false solutions, but also greenwashing[6] (attempt (by companies, institutions) to construct themselves as environmentally aware and friendly through monetary donations for ecological projects, PR measures, etc.).

Just Solutions

Climate change solutions must not only be effective in reducing emissions but also include all, leaving no one behind.

The notion of a *just transition* is at the core of climate justice, wherein workers and their communities receive support during the transition toward a low-carbon economy. The idea of just transition emerged from US labor unions collaborating with environmental justice groups in the 1990s and has now become a significant aspect of global climate conversations. Even the Paris Agreement acknowledges the importance of a *just transition of the workforce* as one of its imperatives.

Ban Fossil Fuels and Corporate Interference

The most effective step to achieve climate justice is to leave fossil fuels in the ground. This also means that neither investments nor public subsidies should be made in fossil-based products. Fossil fuel companies have huge power and their lobbies are strong in our fossil fuel-dependent political and economic systems. Therefore, climate activists have raised concerns about the participation of corporations in UN climate negotiations and have demanded that these companies be held accountable for their prolonged dissemination of misinformation and interference in climate policy. For instance, a report by the NGO Corporate Accountability in 2017 extensively documented how industries with significant greenhouse gas emissions exerted influence on the UN negotiation process. Corporate interference must stop if we are to establish a climate-just society (Gabbatiss & Tandon, 2021).

Overview of This Book

The aim of this book is to present and discuss an integrative (conceptual) model of climate-just behavior and introduce potential solutions to head toward a more climate-just world.

In this introduction, we defined climate (in)justice and climate-just behavior and showed up their historical context. In the following, we investigate existing barriers to as well as opportunities and incentives toward climate-just behavior.

In Chapter 2, we define and describe principles of justice and perception of injustice. Our guiding questions are as follows: How do we perceive injustice? How do we achieve climate justice? In Chapter 3, we take a closer look at human behavior and answer the questions: What factors determine human behavior? What interdependencies and interactions occur? And how can these findings be applied to climate-just behavior?

In Chapter 4, we focus on very specific barriers to climate-just behavior, namely we identify justification mechanisms as a critical barrier to climate-just behavior. We introduce justification mechanisms at individual the level as well as collectively at the societal level.

Finally, we bring our findings together to address the following questions: How can existing individual and collective behavior be transformed into more just climate behavior? What role does the individual play in embedding climate-just behavior in society, political systems and economic dependencies? How do we give up our justification mechanisms to move toward justice?

Notes

1 We are aware that even without climate change, there still would be considerable conflict over water and land. The focus on climate change helps to ignore a world system that for centuries has been based on the exploitation of people and nature. Nevertheless, climate change fosters these mechanisms as it makes water and arable land even more rare.

2 Tool available here: https://berkeleyearth.org/policy-insights/

3 BRICS countries are Brazil, Russia, India, China and South Africa.

4 More precisely, all of this happens independently from climate change but will be exacerbated by climate change.

5 Geoengineering, the deliberate manipulation of the Earth's climate system, has been proposed as a potential solution to counteract global warming (Buck, 2012). This can be achieved through solar radiation management, which aims to reflect more

sunlight to space, or carbon dioxide removal, which aims to reduce CO_2 content in the atmosphere (Cao, 2022). However, the effectiveness and potential impacts of these methods remain under debate (Zhang, 2015).

6 Greenwashing, the deceptive practice of promoting a company's products or services as environmentally friendly when they are not, is a widespread issue across various industries (Aggarwal, 2014). This practice can lead to consumer skepticism and perceived risk, ultimately affecting their trust in green products (Aji, 2015).

References

Aggarwal, P., & Kadyan, A. (2014). Greenwashing: The darker side of CSR. *Indian Journal of Applied Research, 4*(3), 61–66.

Aji, H. M., & Sutikno, B. (2015). The extended consequence of greenwashing: Perceived consumer skepticism. *International Journal of Business and Information, 10*(4), 433.

Ballester, J., Quijal-Zamorano, M., Turrubiates, R. F. M., Pegenaute, F., Herrmann, F., Robine, J., Basagaña, X., Tonne, C., Antó, J. M., & Achebak, H. (2023). Heat-related mortality in Europe during the summer of 2022. *Nature Medicine, 29*(7), 1857–1866. https://doi.org/10.1038/s41591-023-02419-z

Blomfield, M. (2019). *Global justice, natural resources, and climate change.* Oxford University Press eBooks. https://doi.org/10.1093/oso/9780198791737.001.0001

Bressan, L., Cordilha, A. C., Constantino, J. P., & Rubin, P. (2023). *The Brazilian tax system: Regressive and biased.* [Wealth Inequality Working Paper, 54]. Southern Centre for Inequality Studies. https://hdl.handle.net/10539/35647

Brooks, N., Adger, W. N., & Kelly, P. (2005). The determinants of vulnerability and adaptive capacity at the national level and the implications for adaptation. *Global Environmental Change, 15*(2), 151–163. https://doi.org/10.1016/j.gloenvcha.2004.12.006

Buck, H. J. (2012). Geoengineering: Re-making climate for profit or humanitarian intervention? *Development and Change, 43*(1), 253–270. https://doi.org/10.1111/j.1467-7660.2011.01744.x

Cao, Q., Zhou, Y., Du, H., Ren, M., & Zhen, W. (2022). Carbon information disclosure quality, greenwashing behavior, and enterprise value. *Frontiers in Psychology, 13,* 892415.

Chancel, L., Piketty, T., Saez, E., & Zucman, G. (2022). *World inequality report 2022.* World Inequality Lab. https://wir2022.wid.world/www-site/uploads/2023/03/D_FINAL_WIL_RIM_RAPPORT_2303.pdf

Clayton, S., & Karazsia, B. T. (2020). Development and validation of a measure of climate change anxiety. *Journal of Environmental Psychology, 69,* 101434. https://doi.org/10.1016/j.jenvp.2020.101434

Cozzi, L., Chen, O., & Kim, H. (2022, February 15). *What would net zero by 2050 mean for the emissions footprints of younger people versus their parents?* International Energy Agency. https://www.iea.org/commentaries/what-would-net-zero-by-2050-mean-for-the-emissions-footprints-of-younger-people-versus-their-parents

Duus-Otterström, G. (2013). The problem of past emissions and intergenerational debts. *Critical Review of International Social and Political Philosophy, 17*(4), 448–469. https://doi.org/10.1080/13698230.2013.810395

Eckstein, D., Künzel, V., & Schäfer, L. (2021). *Global climate risk index 2021: Who suffers most extreme weather events? Weather-related loss events in 2019 and 2000–2019.* Germanwatch e.V. https://www.germanwatch.org/en/19777

Etchart, L. (2017). The role of Indigenous peoples in combating climate change. *Palgrave Communications*, *3*(1). https://doi.org/10.1057/palcomms.2017.85

Evans, S. (2021, October 4). *Analysis: Which countries are historically responsible for climate change?* Carbon Brief. https://www.carbonbrief.org/analysis-which-countries-are-historically-responsible-for-climate

Gabbatiss, J., & Tandon, A. (2021, October 4). *In-depth Q&A: What is 'climate justice'?* Carbon Brief. https://www.carbonbrief.org/in-depth-qa-what-is-climate-justice/

Gifford, R., & Nilsson, A. (2014). Personal and social factors that influence pro-environmental concern and behaviour: A review. *International Journal of Psychology*, *49*(3), 141–147. https://doi.org/10.1002/ijop.12034

Godard, O. (2012). *Ecological debt and historical responsibility revisited: The case of climate change*. [Working Paper, EUI RSCAS, 2012/46, Global Governance Programme-26, Global Economics]. https://cadmus.eui.eu/handle/1814/23430

Haigh, C., & Vallely, B. (2010). *Gender and the climate change agenda: The impacts of climate change on women and public policy*. Women's Environmental Network (WEN).

Intergovernmental Panel on Climate Change (IPCC). (2022). *Climate change 2022: Impacts, adaption and vulnerability – Summary for policymakers*. Intergovernmental Panel on Climate Change. https://www.ipcc.ch/report/ar6/wg2/downloads/report/IPCC_AR6_WGII_SummaryForPolicymakers.pdf

Jafry, T. (2018). *Routledge handbook of climate justice*. Routledge eBooks. https://doi.org/10.4324/9781315537689

Janmaimool, P., & Watanabe, T. (2012). *Enhancement of disaster preparedness among elderly people by strengthening environmental risk communication*. ResearchGate. https://www.researchgate.net/publication/339886078_enhancement_of_disaster_preparedness_among_elderly_people_by_strengthening_environmental_risk_communication

Klinsky, S., Dowlatabadi, H., & McDaniels, T. L. (2012). Comparing public rationales for justice trade-offs in mitigation and adaptation climate policy dilemmas. *Global Environmental Change*, *22*(4), 862–876. https://doi.org/10.1016/j.gloenvcha.2012.05.008

Kronenberg, M. (2023, February 11). *Erde an Robin Hood: Bitte kommen* [Essay]. Steady. https://steadyhq.com/de/treibhauspost/posts/27b74efd-e13c-4b9e-a7e1-ce75e7030cfc

Layte, R., & Whelan, C. T. (2014). Who feels inferior? A test of the status anxiety hypothesis of social inequalities in health. *European Sociological Review*, *30*(4), 525–535. https://doi.org/10.1093/esr/jcu057

Marks, E., Hickman, C., Pihkala, P., Clayton, S., Lewandowski, E. R., Mayall, E., Wray, B., Mellor, C., & Van Susteren, L. (2021). *Young people's voices on climate anxiety, government betrayal and moral injury: A global phenomenon*. Social Science Research Network. https://doi.org/10.2139/ssrn.3918955

Meikle, M., Wilson, J. F., & Jafry, T. (2016). Climate justice: Between mammon and mother earth. *International Journal of Climate Change Strategies and Management*, *8*(4), 488–504. https://doi.org/10.1108/ijccsm-06-2015-0089

Nelson, V. (2011). *Gender, generations, social protection & climate change: A thematic review*. Overseas Development Institute. https://odi.org/en/publications/gender-generations-social-protection-climate-change-a-thematic-review/

Otto, I. M., Kim, K. M., Dubrovsky, N., & Lucht, W. (2019). Shift the focus from the super-poor to the super-rich. *Nature Climate Change*, *9*(2), 82–84. https://doi.org/10.1038/s41558-019-0402-3

Parfit, D. (2011). *On what matters* (Vol. 1). Oxford University Press.

Pearse, R., & Connell, R. (2015). Gender norms and the economy: Insights from social research. *Feminist Economics, 22*(1), 30–53. https://doi.org/10.1080/13545701.201 5.1078485

Peeters, W., Diependaele, L., & Sterckx, S. (2019). Moral disengagement and the motivational gap in climate change. *Ethical Theory and Moral Practice, 22*(2), 425–447. https://doi.org/10.1007/s10677-019-09995-5

Roser, D., & Seidel, C. (2017). *Climate justice – An introduction* (1st ed.) (C. Cronin, Trans.). Routledge (original published 2013).

Sanson, A., Van Hoorn, J., & Burke, S. (2019). Responding to the impacts of the climate crisis on children and youth. *Child Development Perspectives, 13*(4), 201–207. https://doi.org/10.1111/cdep.12342

Schapper, A. (2018). Climate justice and human rights. *International Relations, 32*(3), 275–295. https://doi.org/10.1177/0047117818782595

Schlosberg, D., & Collins, L. B. (2014). From environmental to climate justice: Climate change and the discourse of environmental justice. *WIREs Climate Change, 5*(3), 359–374. https://doi.org/10.1002/wcc.275

Shaffril, H. A. M., Ahmad, N., Samsuddin, S. F., Samah, A. A., & Hamdan, M. E. (2020). Systematic literature review on adaptation towards climate change impacts among Indigenous people in the Asia Pacific regions, *Journal of Cleaner Production, 258*, 120595. https://doi.org/10.1016/j.jclepro.2020.120595

Shue, H. (1993). Subsistence emissions and luxury emissions. *Law & Policy, 15*(1), 39–60. https://doi.org/10.1111/j.1467-9930.1993.tb00093.x

Shue, H. (2001). Climate. In D. Jamieson (Ed.), *A companion to environmental philosophy* (pp. 449–459). Blackwell.

United Nations (2024). *Indigenous Peoples at the United Nations*. Division for Inclusive Social Development (DISD). Retrieved May 3, 2024, from https://social.desa. un.org/issues/indigenous-peoples/indigenous-peoples-at-the-united-nations.

United Nations Children's Fund (UNICEF). (2023). *The state of the world's children 2023: For every child, vaccination*. UNICEF Innocenti – Global Office of Research and Foresight.

Walasek, L., & Brown, G. D. (2016). Income inequality, income, and internet searches for status goods: A cross-national study of the association between inequality and well-being. *Social Indicators Research, 129*(3), 1001–1014. https://doi.org/10.1007/ s11205-015-1158-4

Watts, N., Amann, M., Arnell, N. W., Ayeb-Karlsson, S., Belesova, K., Berry, H., Bouley, T., Boykoff, M., Byass, P., Cai, W., Campbell-Lendrum, D., Chambers, J., Daly, M., Dasandi, N., Davies, M., Depoux, A., Domínguez-Salas, P., Drummond, P., Ebi, K. L., & Costello, A. (2018). The 2018 report of the Lancet countdown on health and climate change: Shaping the health of nations for centuries to come. *The Lancet, 392*(10163), 2479–2514. https://doi.org/10.1016/s0140-6736 (18)32594-7

Wildcat, D. R. (2013). Introduction: Climate change and Indigenous peoples of the USA. In J.K. Maldonado, B. Colombi, R. Pandya (Eds.), *Climate change and Indigenous peoples in the United States*. Springer. https://doi.org/10.1007/978-3-319-05266-3_1

Williamson, K., Satre-Meloy, A., Velasco, K., & Green, K. (2018). *Climate change needs behavior change: Making the case for behavioral solutions to reduce global warming*. Rare, Center for Behavior & the Environment.

World Bank. (2023, January 17). *The world by income and region*. The World Bank. https://datatopics.worldbank.org/world-development-indicators/the-world-by-income-and-region.html

Wullenkord, M., & Reese, G. (2021). Avoidance, rationalization, and denial: Defensive self-protection in the face of climate change negatively predicts pro-environmental behavior. *Journal of Environmental Psychology, 77*, 101683. https://doi.org/10.1016/j.jenvp.2021.101683

Zhang, Z., Moore, J. C., Huisingh, D., & Zhao, Y. (2015). Review of geoengineering approaches to mitigating climate change. *Journal of Cleaner Production, 103*, 898–907.

2 A Psychological Perspective on Justice and Injustice

Susanne Nicolai

Relevance of Justice

As the opposite of injustice, justice is an ideal state of balanced interests without the disadvantage of individuals or groups. It is regarded in all cultures as a central value and imperative of individual and collective action. Contrary to injustice, justice is difficult to recognize in practice. When considering climate (in)justice, a critical question arises: What makes humans perceive a situation as just or unjust?

First of all, justice has a high value for people. It provides orientation in an overcomplex world (Müller, 2014), counts as a motivator (justice motive, Lerner, 1975) and helps regulate and maintain social relationships (Gollwitzer & Van Prooijen, 2016). Justice is a relevant part of social relationships and is a basis for stable societies, as it acts as a "social glue" that significantly contributes to the effective functioning of organizations (or societies). Justice is considered the highest principle in justifying normative order, one of the most important virtues of our society, and is always connected with the idea of the good (Peter, 2012). To maintain justice, people are willing to accept financial losses (Engel, 2011), not enter a profitable deal (Sutter et al., 2019) or change their behavior (Kals & Becker, 2006). The desire for justice stems from the need for security and control over one's life. If the world is experienced as just, the "rules of the game" are clear, transparent and can be anticipated (Lerner, 1980; Dalbert & Katona-Sallay, 1996).

Justice Principles

Justice has many dimensions: (1) *interactional justice* focuses on issues associated with interpersonal interactions, (2) *procedural justice* relates to official processes (e.g. hiring employees) (3) *restorative justice* deals with compensation/reparation after injuries or crimes and (4) *distributive justice* encompasses aspects related to the allocation of resources. In order to assess a situation as just or unjust, various principles of justice can be defined

DOI: 10.4324/9781003033547-2

according to the situation at hand. In the sustainability field, distributive justice prevails, that is, on issues on how to justly distribute resources as well as burdens (CO_2 budgets) to meet the 1.5°C limit.

The assessment of whether all parties have received their fair share can proceed along three different principles, including of equality (i.e. everyone receives the same amount) of merit (i.e. everyone receives as much as they contributed to through their performance), and of need (i.e. everyone receives as much as they need) (Ehrhardt-Madapathi et al., 2018; Jasso et al., 2016). Depending on the context, situation, person and relationship, each of these principles may be perceived as having different levels of justice and appropriateness (Deutsch, 1985; Jasso et al., 2016; Skitka et al., 2016). Furthermore, there might be contradictions and trade-offs between these principles and dimensions of justice. In terms of the context of climate justice, individuals differ on which principles they prioritize in evaluating the distribution of resources and tasks, and how.

The Intergovernmental Panel on Climate Change (IPCC) addresses these points by highlighting three principles of climate justice in their AR6 (2022): distributive justice, procedural justice and recognition. (1) Distributive justice in the climate domain refers to the allocation of burdens and benefits among individuals, nations and generations; (2) procedural justice refers to who decides and participates in decision-making and (3) recognition entails basic respect and robust engagement with, and fair consideration of, diverse cultures (and in particular indigenous knowledge) and can therefore be associated with interactional justice. Nonetheless, the current recognition status quo is incredibly unjust: for example, most politicians, members of the management boards of DAX companies as well as even climate scientists are still white, male, middle aged or older and affluent.

Justice Theories

One of the most influential theories in the scientific literature about justice is the equity theory by Adams (1965), which states that "just" means "equal" when assessing the distribution of resources. To be equal, according to Adams, means having the same ratio of inputs to outputs as all other people. In the case of inequality, people have the chance to either reframe the situation as equal (cognitive reconstruction) or become active to change the situation to restore justice. However, this approach is criticized for its focus on the principle of merit (i.e. the more input the more output) over other justice principles. Furthermore, it is unclear with whom comparisons are made and how inputs and outputs are weighted (Peter, 2012).

In this theory in the context of climate justice the ration between inputs and outputs could put forward the polluter-pays-principle as a just solution. However, Adams's theory is embedded within a neo-liberal worldview in which performance alone is rewarded, while obstacles to performance are ignored.

Keeping climate injustice in mind, vulnerable groups may need higher outputs (e.g. money) to adapt to climate change consequences.

In our opinion, a more useful theory is the *Referent Cognition Theory* by Folger (1986), which is a response to the equity theory. Folger contends that injustice goes along with negative emotions due to the discrepancy between perceived reality and imaginary alternatives. Folger's theory is not only adaptable to distributive justice but also to interactional and procedural justice. Moreover, it does not concentrate on the principle of merit only but also regards the principles of equality and need (Peter, 2012). Thus, in a study on pro-environmental behavior, we demonstrated that negative emotions (e.g. guilt) in keeping with others are closely related to the perception of injustice. These negative emotions can be used to predict pro-environmental intention and behavior and can be a way to restore justice (Nicolai et al., 2022). We will discuss the role of emotions in Chapter 3 on behavior. In the following, we describe that injustice can motivate (pro-environmental) behavior.

Justice as a Motivational Factor and Fundamental Motive

Justice has been acknowledged a fundamental human motive (Baumert et al., 2013a; Montada, 2007). Consequently, individuals generally seek to establish justice and avoid injustice (Lerner, 1977). Further, individuals want themselves and others to be treated fairly and are willing to behave according to justice principles (Baumert et al., 2013b).

Baumert et al. (2013b) summarize:

> assuming that justice is a fundamental motive for individuals means that the perception of a potential injustice triggers emotional reactions (e.g. anger, moral outrage, compassion, guilt) and urges the individual to act in order to restore justice or to avoid the injustice. Hence, the concept of a human justice motive implies the assumption of a psychological link between the perception of (potential) injustice and affective and behavioral reactions.
>
> (p. 161)

And indeed, perceived (in)justice or fairness is one major predictor of both, pro-environmental behavior and acceptance of (mitigation and adaptation) policies (Clayton, 2018; Kals & Becker, 2006; Nicolai et al., 2022). For example, in a study by Clayton (2018) ratings of the fairness of specific policies were a stronger determinant of acceptability than perceived effectiveness of the policy. Furthermore, a broad endorsement of environmental justice was found to have a significant association with the acceptance of all listed climate policy strategies in this study (e.g. prohibitions, incentives, business tax or carbon credit), often surpassing the influence of political orientation and consistently outweighing the impact of environmentalism.

Moreover, to establish justice, individuals are willing to give up their own advantages (Engel, 2011), act sustainably (Kals & Becker, 2006) or reject a beneficial but unfair deal (Sutter et al., 2019). A meta-analysis based on 182 studies on collective behavior confirmed that the more strongly that individuals perceive a situation to be unjust, the more likely they are to participate in collective action (such as protest behavior) to improve the situation (Van Zomeren et al., 2008).

Although justice is a fundamental motive, individuals differ in their judgments concerning whether a specific situation (e.g. the climate crisis) is just or unjust. Therefore, it is important to understand what mechanisms underlie this different perception of injustice. These differing assessments have been related to diverge principles of justice that vary across individuals, contexts, situations and relationships (Deutsch, 1985; Jasso et al., 2016; Sabbagh & Schmitt, 2016; Skitka et al., 2016).

Perceived Injustice and Justice Sensitivity

Individuals systematically differ in terms of how strongly they perceive injustice. These differences in the intensity of injustice perception, as well as associated reactions in terms of emotions and behavior, are summarized under the term injustice sensitivity (Baumert & Schmitt, 2016; Schmitt et al., 1995, 2005). Injustice sensitivity is stable across situations, contexts and time and is therefore seen as a personality trait (Schmitt et al., 1995, 2005). However, different facets occur not only between but also within a person: injustice sensitivity can be perceived differently from perpetrator, victim, beneficiary and observer perspectives. Even though justice is a concern for all perspectives, striking differences emerge. For example, persons with high values for observer, beneficiary and perpetrator sensitivity show a strong striving for justice for others (not oneself!) as well as a sense of social responsibility. On the other hand, high scores on the victim sensitivity dimension are associated with a drive for justice for oneself (Preiser & Beierlein, 2017).

Furthermore, these differences are reflected in the emotions that arise: individuals with high perpetrator and beneficiary sensitivity tend to feel more guilt than those with lower levels. Observer-sensitive people mainly feel moral outrage when they observe injustice, and victim-sensitive people mostly feel anger when experiencing injustice (Preiser & Beierlein, 2017). This in turn results in different perceptions of how to restore the justice that has been violated. Perpetrator and beneficiary sensitives seek to make amends, victim sensitives often possess a strong desire to punish the perpetrators rather than change their own behavior (Gollwitzer et al., 2009; Preiser & Beierlein, 2017; Schmitt et al., 2009; review: Landmann, 2020). Thus, the perspective from which people perceive injustice, strongly influences the emotions associated with injustice, which, in turn, generates a specific tendency for action. The relevant difference is the perception of injustice and accompanying emotional facets.

In a recent study, we were able to show that pro-social injustice sensitivities significantly and positively predicted pro-environmental intentions. Victim sensitivity did predict pro-environmental intention negatively, and the tendency to morally disengage in high-carbon behavior, positively (Nicolai et al., 2022). The prediction of perpetrator sensitivity and beneficiary sensitivity on pro-environmental intention was mediated by the moral emotion of guilt. Therefore, it seems promising to make the injustice of climate change more salient and foster pro-social justice sensitivities in society.

A first study in this field (Maltese et al., 2013) shows that specific ambiguity training can teach participants to classify the consequences of their own behavior as unjust. Applied to the climate crisis, the aim of what is to draw attention onto why the climate crisis is fundamentally a question of justice and how one's own behavior is connected to the unjust consequences of the climate crisis. This can lead to collective climate anger, which can provide the impetus to demand just distribution of climate protection measures and resources for adaptation politically and socially and thus make it possible. For example, Traut-Mattausch et al. (2011) state that the perception of injustice is the necessary starting point for any social protest. Moreover, social protest or collective action is an opportunity to change the established system (as we will further explain in Chapter 3 with the multilevel model).

Requirements to Perceive Injustice

In the context of climate justice, people in industrialized countries can be regarded as the beneficiaries of the detrimental living conditions of people from low-income countries, since the standard of living in industrialized countries is directly connected to the exploitation of natural and human resources low-income countries (Dorninger et al., 2021; Hickel et al., 2022). However, many people may not be completely aware of this fact as the actions that result in driving climate change are often perceived as an "unintentional, if unfortunate, side effect" of goal-directed behavior (Markowitz & Shariff, 2012). Markowitz and Shariff (2012) call this phenomenon the blamelessness of unintentional action. Unintentionally caused consequences are judged less harshly than equally severe but intentionally caused consequences (Guglielmo et al., 2009). However, the existence or lack of an intention does not change the consequences of an action. Linking one's own behavior with their distant socio-environmental consequences requires some knowledge, immunity from corporate greenwashing discourse and the disposition to perceive the injustice they are contributing to from the perspective of beneficiaries, perpetrators or observers. Therefore, we hypothesize that people need to acknowledge their responsibility to renounce their privileges and act in solidarity with the disadvantaged. Moreover, we assume that this perception is a motivator for pro-environmental behavior change and, therefore, is pivotal to achieving climate justice.

However, a sense of empathy seems to be important to fully understand the situation of the most affected groups. People not only need to perceive justice cognitively but also must be emotionally motivated by compassion or a genuine concern for the well-being of others to become active. For example, Roeser (2012) found that emotional competencies are necessary to understand the moral dimension of a specific decision. The author confirms that emotions may help to perceive something that purely rational assessments fail to discover. In her study, she demonstrated that information about climate change associated with emotional content provided readers better insight into its moral meaning as well as a deeper, more reliable source of motivation for action than information without. We contend that cognitive and behavioral empathy, that is, the ability to correctly recognize the feelings and thoughts of others as well as to react appropriately to them, should therefore also be promoted.

Inaction as Injustice

The term "passive injustice" coined by Shklar's (2021) takes into account the fact that injustice does not always have to be the consequence of an active action but can also come into the world through omission: anyone who can do something to prevent injustice or mitigate the extent of its consequences and does not act, although is behaving in a passively unjust way. This was the verdict of the German Federal Constitutional Court in its landmark ruling of 2021, which qualified German climate protection law insufficient to guarantee the freedoms of future generations. Indeed, the efforts of the German government to protect citizens from climate change impacts were classified as insufficient on the argument of failing to properly act to counter climate impacts in the face of available evidence.

What counts as passive injustice is often defined on a case-by-case basis, although they may be strongly associated to one's plausible expectations of state actors and fellow citizens in a liberal democracy, such as being protected from disasters or receiving help after being impacted by such disasters. However, Shklar understands passive injustice not so much as a moral category but rather primarily political, as a form of unfulfilled democratic expectations. Such expectations affect private behavior, given that we expect our fellow citizens not to look the other way when injustice happens to us. They likewise concern state representatives, as they too can be passively unjust when they default and neglect their official duties, delay action or mismanage. For Shklar, the goal of preventing passive injustice is not exhausted in the search for the guilty, as scarce resources should be prioritized on remedying the suffering of the victims.

However, in the field of climate justice, it is necessary to identify origins and causers (high emitters), as they need to stop continuing. Those responsible for climate change did not do so in one single event but in continuous actions that emit lots of greenhouse gasses. Therefore, one part is to appeal to the responsibility of the polluters to change their high-emitting lifestyles (mitigation) and, second – and in keeping with Shklar – to give support to the most vulnerable people (adaptation).

Individual Behavior in an Unjust World

In this chapter, we have shown that justice is a cross-cultural, fundamental human motive. On the other hand, people are confronted with considerable climate injustice. According to the Referent Cognition Theory (see above), this discrepancy leads to negative emotions. In Chapter 4, we will demonstrate that justification can be a coping mechanism to deal with these negative emotions based on the theory of cognitive dissonance and moral disengagement.

Nonetheless, justice is part of social systems and therefore can vary between social systems (or when a social system is changing). Depending on the context, situation, person and relationship, different principles of justice are considered appropriate. This appropriateness may develop via social systems. In a climate-unjust world, we grow up being told that injustices such as the profound gap between the poor and the rich are justified or at least explainable. Pervasive neoliberalism gives the illusion that everyone can become rich, and that wealth is the result of hard work, a fact that based on the principle of merit is perceived as just. Moreover, social norms apply: differences in countries' approaches to justice exist, which also means that the collective perception of justice can be changed. Potentially targets for change could be working to deconstruct justifications of obvious climate-damaging behavior to make visible their inherent injustice. Individual perception and behavior therefore are interrelated with collective perception and behavior, which we will take a closer look at in the next chapter. We understand perceived injustice as a bridge from social embedding to the individual (e.g. in how far the political and economic system influences us) but also the other way around from the individual to its social contexts (e.g. collective behavior like demonstrations or petitions due to perceived injustice).

As the IPCC (2022) concludes, to become more climate resilient, our societies need to become more equal and just.

> Climate resilient development is enabled when governments, civil society and the private sector make inclusive development choices that prioritize risk reduction, **equity and justice**, and when decision-making processes, finance and actions are integrated across governance levels, sectors and timeframes (very high confidence). Climate resilient development is facilitated by international cooperation and by governments at all levels working with communities, civil society, educational bodies, scientific and other institutions, media, investors and businesses; and by **developing partnerships with traditionally marginalized groups, including**

women, youth, Indigenous Peoples, local communities and ethnic minorities (high confidence). These partnerships are most effective when supported by enabling political leadership, institutions, resources, including finance, as well as climate services, information and decision support tools (high confidence).

(IPCC, 2022, p. 14; emphases added by
the authors of the IPCC report)

References

Adams, J. S. (1965). Inequity in social exchange. In L. Berkowitz (Ed.), *Advances in experimental social psychology* (pp. 267–299). Academic Press. https://doi.org/10.1016/s0065-2601(08)60108-2

Baumert, A., & Schmitt, M. (2016). Justice sensitivity. In C. Sabbagh & M. Schmitt (Eds.), *Handbook of social justice theory and research* (pp. 161–180). Springer. https://doi.org/10.1007/978-1-4939-3216-0_9

Baumert, A., Halmburger, A., & Schmitt, M. (2013a). Interventions against norm violations: Dispositional determinants of self-reported and real moral courage. *Personality and Social Psychology Bulletin, 39*(8), 1053–1068. https://doi.org/10.1177/0146167213490032

Baumert, A., Rothmund, T., Thomas, N., Gollwitzer, M., & Schmitt, M. (2013b). Justice as a moral motive: Belief in a just world and justice sensitivity as potential indicators of the justice motive. In K. Heinrichs, F. Oser, & T. Lovat (Eds.), *Handbook of moral motivation: Theories, models, applications* (pp. 159–179). Sage.

Clayton, S. (2018). The role of perceived justice, political ideology, and individual or collective framing in support for environmental policies. *Social Justice Research, 31*(3), 219–237. https://doi.org/10.1007/s11211-018-0303-z

Dalbert, C., & Katona-Sallay, H. (1996). The "Belief in a just world" construct in Hungary. *Journal of Cross-Cultural Psychology, 27*(3), 293–314. https://doi.org/10.1177/0022022196273003

Deutsch, M. (1985). *Distributive justice: A social-psychological perspective*. Yale University Press.

Dorninger, C., Hornborg, A., Abson, D. J., Von Wehrden, H., Schaffartzik, A., Giljum, S., Engler, J., Feller, R. L., Hubacek, K., & Wieland, H. (2021). Global patterns of ecologically unequal exchange: Implications for sustainability in the 21st century. *Ecological Economics, 179*, 106824. https://doi.org/10.1016/j.ecolecon.2020.106824

Ehrhardt-Madapathi, N., Bohndick, C., Holfelder, A. K., & Schmitt, M. (2018). *Nachhaltigkeit in primären und sekundären Bildungseinrichtungen* (pp. 57–64). Springer eBooks. https://doi.org/10.1007/978-3-658-19965-4_5

Engel, C. (2011). Dictator games: A meta study. *Experimental Economics, 14*(4), 583–610. https://doi.org/10.1007/s10683-011-9283-7

Folger, R. (1986). Rethinking equity theory: A referent cognitions model. *Justice in social relations* (pp. 145–162). Springer.

Gollwitzer, M., & Van Prooijen, J. (2016). Psychology of justice. *Handbook of social justice theory and research* (pp. 61–82). Springer eBooks. https://doi.org/10.1007/978-1-4939-3216-0_4

Gollwitzer, M., Rothmund, T., Pfeiffer, A., & Ensenbach, C. (2009). Why and when justice sensitivity leads to pro- and antisocial behavior. *Journal of Research in Personality, 43*(6), 999–1005. https://doi.org/10.1016/j.jrp.2009.07.003

Guglielmo, S., Monroe, A. E., & Malle, B. F. (2009). At the heart of morality lies folk psychology. *Inquiry: An Interdisciplinary Journal of Philosophy, 52*(5), 449–466. https://doi.org/10.1080/00201740903302600

Hickel, J., Dorninger, C., Wieland, H., & Suwandi, I. (2022). Imperialist appropriation in the world economy: Drain from the global south through unequal exchange, 1990–2015. *Global Environmental Change, 73,* 102467. https://doi.org/10.1016/j.gloenvcha.2022.102467

Intergovernmental Panel on Climate Change (IPCC). (2022). *Climate change 2022: Impacts, adaption and vulnerability – Summary for policymakers.* Intergovernmental Panel on Climate Change. https://www.ipcc.ch/report/ar6/wg2/downloads/report/IPCC_AR6_WGII_SummaryForPolicymakers.pdf

Jasso, G., Törnblom, K. Y., & Sabbagh, C. (2016). Distributive justice. *Handbook of social justice theory and research* (pp. 201–218). Springer.

Kals, E., & Becker, R. P. (2006). *Zusammenschau von drei umweltpsychologischen Untersuchungen zur Erklärung Verkehrsbezogener Verbotsforderungen, Engagementbereitschaften und Handlungsentscheidungen. Berichte aus der Arbeitsgruppe "Verantwortung, Gerechtigkeit, Moral": Bd* (p. 73). Universitäts- und Landesbibliothek.

Landmann, H. (2020). Emotions in the context of environmental protection: Theoretical considerations concerning emotion types, eliciting processes, and affect generalization. *Umweltpsychologie, 24*(2), 61–73. http://umps.de/php/artikeldetails.php?id=745

Lerner, M. J. (1975). The justice motive in social behavior: Introduction. *Journal of Social Issues, 31*(3), 1–19. https://doi.org/10.1111/j.1540-4560.1975.tb00995.x

Lerner, M. J. (1977). The justice motive: Some hypotheses as to its origins and forms. *Journal of Personality, 45*(1), 1–52. https://doi.org/10.1111/j.1467-6494.1977.tb00591.x

Lerner, M. J. (1980). *Belief in a just world: A fundamental delusion.* Plenum Publishing Corporation.

Maltese, S., Baumert, A., Knab, N., & Schmitt, M. (2013). Learning to interpret one's own outcome as unjustified amplifies altruistic compensation: A training study. *Frontiers in Psychology, 4* https://doi.org/10.3389/fpsyg.2013.00951

Markowitz, E. M., & Shariff, A. F. (2012). Climate change and moral judgement. *Nature Climate Change, 2*(4), 243–247. https://doi.org/10.1038/nclimate1378

Montada, L. (2007). *Justice conflicts and the justice of conflict resolution.* In *Distributive and procedural justice: Research and social applications* (pp. 255–268). Routledge.

Müller, W. (2014). Educational inequality and social justice: Challenges for career guidance. In G. Arulmani, A. Bakshi, F. Leong, & A. Watts (Eds), *Handbook of career development. International and cultural psychology* (pp. 335–355). Springer. https://doi.org/10.1007/978-1-4614-9460-7_19

Nicolai, S., Franikowski, P., & Stoll-Kleemann, S. (2022). Predicting pro-environmental intention and behavior based on justice sensitivity, moral disengagement, and moral emotions – Results of two quota-sampling surveys. *Frontiers in Psychology, 13.* https://doi.org/10.3389/fpsyg.2022.914366

Peter, F. (2012). Die Bedeutung intuitiver Gerechtigkeitsvorstellungen für Schülerinnen und Schüler: Eine mehrebenenanalytische Längsschnittuntersuchung zur

Wechselwirkung von implizitem Gerechtigkeitsmotiv und schulischer Umwelt. *Schriften zur pädagogischen Psychologie* (p. 57). Verlag Dr. Kovac.

Preiser, S., & Beierlein, C. (2017). Gerechtigkeitsempfinden. In J. Sautermeister (Ed.), *Moralpsychologie: Transdisziplinäre Perspektiven* (1st ed., pp. 262–274). Kohlhammer.

Roeser, S. (2012). Risk communication, public engagement, and climate change: A role for emotions. *Risk Analysis, 32*(6), 1033–1040. https://doi.org/10.1111/j.1539-6924.2012.01812.x

Sabbagh, C., & Schmitt, M. (2016). *Handbook of social justice theory and research.* Springer eBooks. https://doi.org/10.1007/978-1-4939-3216-0

Schmitt, M., Neumann, R., & Montada, L. (1995). Dispositional sensitivity to befallen injustice. *Social Justice Research, 8*(4), 385–407. https://doi.org/10.1007/bf02334713

Schmitt, M., Gollwitzer, M., & Arbach, D. (2005). Justice sensitivity. *European Journal of Psychological Assessment, 21*(3), 202–211. https://doi.org/10.1027/1015-5759.21.3.202

Schmitt, M., Baumert, A., Fetchenhauer, D., Gollwitzer, M., Rothmund, T., & Schlösser, T. (2009). Sensibilität für Ungerechtigkeit. *Psychologische Rundschau, 60*(1), 8–22. https://doi.org/10.1026/0033-3042.60.1.8

Shklar, J. N. (2021). *Über Ungerechtigkeit. Erkundungen zu einem moralischen Gefühl.* Matthes & Seitz.

Skitka, L. J., Bauman, C. W., & Mullen, E. (2016). Morality and justice. In C. Sabbagh & M. Schmitt (Eds.), *Handbook of social justice theory and research* (pp. 407–423). Springer eBooks. https://doi.org/10.1007/978-1-4939-3216-0_22

Sutter, M., Huber, J., Kirchler, M., Stefan, M., & Walzl, M. (2019). Where to look for the morals in markets? *Experimental Economics, 23*(1), 30–52. https://doi.org/10.1007/s10683-019-09608-z

Traut-Mattausch, E., Guter, S., Zanna, M. P., Jonas, E., & Frey, D. (2011). When citizens fight back: Justice sensitivity and resistance to political reform. *Social Justice Research, 24*(1), 25–42. https://doi.org/10.1007/s11211-011-0125-8

Van Zomeren, M., Postmes, T., & Spears, R. (2008). Toward an integrative social identity model of collective action: A quantitative research synthesis of three sociopsychological perspectives. *Psychological Bulletin, 134*(4), 504–535. https://doi.org/10.1037/0033-2909.134.4.504

3 The Complexity of Human Behavior and What This Means for Explaining Climate-Related Behavior

Susanne Stoll-Kleemann

Understanding why people do what they do is central to advancing equitable and sustainable futures.

(Eyster et al., 2022, p. 725)

Human behavior is complex and is shaped by many influences (Roth, 2016; Sapolsky, 2017). This principle also applies to climate-related behavior. Various disciplines, theories and views offer valuable insights into the reasons that drive human behavior. The complexity of human motivation is influenced by individual characteristics, personal experiences, situational and external factors and their interplay. People can be motivated by different combinations of these factors, and motivations can also evolve over time (Eyster et al., 2022; Roth, 2016; Sapolsky, 2017; Stoll-Kleemann, 2019).

Given these complex relationships, it is unsurprising that there is no comprehensive theory of human behavior. Instead, there are numerous theories from various fields such as neuroscience, social and personality psychology, sociology, economics and political science, each of which explains only certain aspects of behavior. It is currently generally accepted that human behavior is shaped by a combination of internal variables such as genetics, personality traits, cognitive processes and psychological needs, as well as external influences such as the political-economic system, the environment and culture. The debate on "nature versus nurture", which examines the influence of individual characteristics and external factors on human behavior, has been settled due to extensive research showing the importance of both areas and their interconnectedness (Eyster et al., 2022; Roth, 2016; Sapolsky, 2017).

In the first chapters of this book, it is explicitly highlighted that influencing social and environmental conditions, such as climate justice, requires a change in both individual and collective human behavior, especially among "the rich". Unfortunately, there is a conflict between the complexity of realizing a sustainable future, which is characterized by numerous interlinked contexts of human action, and the limited availability of comprehensive theories (Eyster et al., 2022). Current theories of human behavior are too

DOI: 10.4324/9781003033547-3

dispersed across numerous scientific disciplines, especially in the social sciences. Consequently, they are unable to effectively address the various issues related to climate-related behavior. Therefore, a comprehensive approach to understanding human behavior is essential, which can shed light on the reasons for individuals' actions and the potential for change. Furthermore, it is crucial to incorporate often neglected but important theories of collective action and structural change (Eyster et al., 2022).

We provide a brief overview of the efforts of Eyster et al. (2022) to combine 86 theories from different fields, mainly from the social sciences, with the aim of promoting feasible behavioral (and societal) change to achieve climate justice. Eight main "meta-theories" are distinguished, which are understood as the quintessence of many individual behavioral theories that are combined and abstracted. Each assumes that a unique combination of elements is responsible for the emergence of human behavior and enables the understanding of a particular aspect, extent and cause of human action (see Figure 3.1). For example, the meta-theory of the "independent self" assumes that individual behavior is influenced by personal characteristics such as knowledge, values, attitudes, character traits and (religious) beliefs, which have a unidirectional effect on behavior. In contrast, the "independent structure" meta-theory assumes that human behavior is primarily influenced by autonomous structural elements such as culture, education, learning environments, institutions and infrastructure (Eyster et al., 2022). Both meta-theories do not take into account feedback, such as positive feedback loops, or interactions between structural variables and human characteristics.

	Metatheory	What does it enable?	Scale of analysis	Scale of change	Lever identification	Applicable population	Implementation speed	Example application to address climate change
Independent	Self	Change individual attitudes toward decisions (structure/context is constant)	Small	Small, incremental	Usually	Homogeneous		Create positive individual attitudes toward bike commuting
	Structure	Change institutions to enable change, holding individuals constant	Medium	Medium, moderately incremental	Usually	Heterogeneous	Variable	Create separated bike lanes; build bike cages
Needs	Cognitive	Change cues to harmonize cognitive needs with specific choice (e.g., nudges)	Small	Medium, incremental	Usually	Heterogeneous	Rapid, cheap	Require parking passes to be manually renewed
	Psychological	Intrinsically motivate people, which in turn increases subjective well-being	Small	Medium, moderately systemic	Usually	Heterogeneous	Variable	Redesign institutions to enable people to meet their need for consonance between action and underlying environmental values
	Communal	Create adaptive, communal, equitable processes	Medium	Medium, moderately systemic	Usually	Heterogeneous	Slow	Change the organizational structure and processes of a climate activism organization
	Economic	Change costs of choices within a static, homogeneous system	Small	Medium, incremental	Always	Homogeneous	Rapid	Introduce a carbon tax
Interdependent		Identify and intervene in dynamic feedbacks that (re)produce social practices	Large	Large, systemic, and adaptive	Sometimes	Heterogeneous	Slow	Change competence, availability, meanings, and technology about biking
Top-down		Identify overarching problem	Large	Large, systemic	Rarely	Heterogeneous	Slow	Reorganize nation's political economy

Figure 3.1 Key attributes of Eyster et al.'s (2022) meta-theories that inform suitable application and an example solution each might propose to tackle climate change

The "cognitive needs meta-theory" states that the fundamental goal of human behavior is "survival", which is achieved through the efficient cognitive processing of information in conjunction with heuristic processing (Eyster et al., 2022). This implies the involvement of unconscious, rapid cognitive processing that prioritizes existing knowledge and peripheral information, sometimes referred to as "fast thinking" (see below; Chaiken, 1980; Evans, 2008 in Eyster et al., 2022; Kahneman, 2011).

The "psychological needs meta-theory" assumes that the primary goal of human behavior is to generate subjective well-being. Eyster et al. (2022) identified six psychological needs to be satisfied, including the promotion of pleasure, avoidance of pain, autonomy, competence, consistency and relatedness. The knowledge base provided by this meta-theory includes several foundational theories such as self-determination theory, social norms, cognitive dissonance, social-cognitive theory and self-efficacy. The insights from these theories can generally facilitate the implementation of strategies for short-term and long-term change. In addition, the theories suggest that by using the needs for relatedness and competence, it is possible to gradually align behavior with social norms (Cialdini et al., 2006; Naito et al., 2022; Ryan & Deci, 2000 in Eyster et al., 2022). Furthermore, psychological needs theories can effectively influence the behavior of heterogeneous populations due to their universal nature (see Figure 3.2; Eyster et al., 2022). The promotion

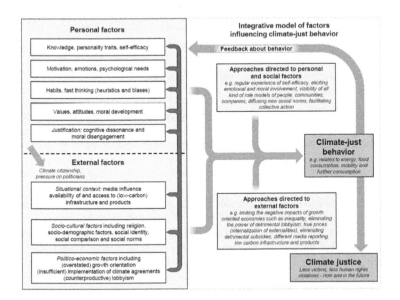

Figure 3.2 Integrative Model of Factors influencing Climate-Just Behavior (Strongly modified after Stoll-Kleemann 2019)

of connectedness and pleasure can be used to support leaders who can bring about significant change in environmental movements.

The basic principles of the "communal needs meta-theory" focus on co-operation, for example, social cooperation, collective action and effective governance, as well as the fulfillment of community needs within specific institutions and cultures. Nevertheless, established power hierarchies can hinder the fulfillment of collective needs or limit the influence of these groups (Eyster et al., 2022). The "top-down meta-theory" assumes that human behavior and personal characteristics are primarily influenced by higher level and systemic factors such as "culture, beliefs, economic and political systems" in a one-sided manner. This meta-theory has already proven useful in identifying the underlying systemic problems "that have generated the current climate crisis and impeded its resolution such as nature domination, instrumental use, endless growth, and corporate power that undergird our current political-economic system" (Eyster et al., 2022, p. 740; Martínez-Alier, 2014).

Eyster et al. (2022) have concluded that the "interdependent meta-theory" is the most appropriate approach to tackling complex, unpredictable and interconnected challenges such as the climate crisis. This theory assumes that human action is "continually created, reinforced, or eroded by an interdependent web of values, identities, positions, habits, goals, needs, experiences, meanings, institutions, cultures, politics, etc." (Eyster et al., 2022, p. 741). It is evident that effective transformative strategies to mitigate the climate crisis, which is a highly complex global problem, require a strategic integration of multiple complementary approaches (Meadows, 2008; O'Connor et al., 2021 in Eyster et al., 2022).

A Conceptual Integrative Model of Climate (Un)Just Behavior

This conclusion is a call to further develop an existing model that has been applied to three different climate-related behaviors (meat consumption, sustainable ocean behavior and sustainable agricultural practices) by proposing an improved, more integrative model framework for climate (un)just behavior based on new research findings. This model already includes feedback to show nonlinear and complex relationships within and between human behavior as well as sociocultural and political-economic factors and other external circumstances related to the climate crisis. All three only slightly different versions of a model are based on extensive meta-analyses or first-hand interviews and have been published in scientific journals (Schwerdtner Máñez et al., 2024; Stoll-Kleemann, 2019; Stoll-Kleemann & Schmidt, 2017). All versions are based on the model of environmentally friendly behavior developed by Kollmuss and Agyeman (2002), which meets the important requirements of interdependence and feedback. The decision to use this model as the basis for creating a new integrative model was informed by a review of numerous theoretical frameworks of individual behavior, drawing on Darnton's (2008) comprehensive analysis of behavior change models. The main reason for choosing and using this model to explain climate-related behaviors is primarily due to its comprehensive scope and multifaceted approach. Gifford and Nilsson (2014) argue that several studies have pointed to the need to extend established social-psychological models such as the theory of planned behavior (Ajzen, 1991), the value-belief-norm model (Stern, 2000) and the theory of norm activation (Schwartz, 1977) by including additional personal and social factors. This finding is also supported by Eyster et al. (2022) in relation to theories that emphasize only top-down and/or independent structural factors.

What is new about this fourth version of the model is the explicit thematic focus on climate-just behavior and the inclusion of other and new literature, such as findings from neuroscience. In particular, external factors such as the role of the media and climate lobbying and an excessive economic growth orientation in our current capitalist system are given greater consideration. Finally, we add useful new empirical findings from our representative studies on climate justice issues.

The model presented here is divided into personal and external factors (including sociocultural factors). For the selection of personal factors, the influences of knowledge, personality traits, self-efficacy, moral development, motivations, psychological needs, values, attitudes and moral development as well as habits and cognitive processes such as quick thinking on potential climate justice behavior were examined. Emotions (as a central component of motivational systems) have an overwhelming influence on several of the other model components, such as moral disengagement, which is discussed in detail in Chapter 4. Emotions and moral disengagement are core factors for climate-just behavior because they help us to understand the personal reluctance to reduce or avoid climate-just behavior that has been discussed by one of the authors for more than 20 years (see e.g. Stoll-Kleemann et al., 2001). The arrows in the model show how the various factors influence each other and ultimately climate-just behavior. On the right-hand side of the figure, a distinction is made between different approaches that focus more on the personal and social factors of behavior on the one hand and on the more external factors on the other. These approaches and interventions should be supported by rewarding feedback on those behaviors that lead to less harm for people suffering from the climate crisis.

Personal Factors

Knowledge, Personality Traits and Self-efficacy

Sufficient knowledge is an important but overestimated precondition for achieving climate-just behavior. For the general public, there is no evidence that more knowledge, education and public awareness of the climate crisis leads to climate-just behavior. Contrary to that popular belief, studies indicate that "only a small fraction of pro-environmental behavior can be directly linked to environmental knowledge and environmental awareness" (Kollmuss & Agyeman, 2002, p. 250). The gap between knowledge/awareness and behavior – which can be found in all areas of life – is explained by the fact that "at least 80% of the motives for pro-environmental or non-environmental behavior seem to be situational factors and other internal factors" (Barr et al., 2011; Barreto et al., 2014; Kollmuss & Agyeman, 2002, p. 250). More specifically, this implies that while individuals may acknowledge their concerns, a majority of them ultimately pay less attention to climate-just behavior in their day-to-day actions and decisions. Furthermore, it is important to consider that every behavior is linked with subjective behavioral costs which can be perceived as being either high (e.g. intolerably prolonged travel time) or low and simple (e.g. resetting the default of a thermostat or switching off lights when leaving a room) (Rau et al., 2024; Diekmann & Preisendorfer, 2003).

However, in addition to factual knowledge (in the sense of knowledge of issues), procedural knowledge (skills in the sense of knowledge of action strategies) is very important. This "skills" component is underestimated but should be taken seriously. It is often "acquired through experience or observation, as much as through formal information" (Darnton & Evans, 2013, p. 15). Regarding the skills component, there is also more provision of concrete low-carbon activities in the mass as well as in the social media. With several cook shows knowledge of "meat-free" recipes increases and the former lack of skills regarding how to use meat substitutes in cooking decreases. A further study investigates the group of farmers: a review of 279 studies on the adoption of agricultural conservation practices in the United States identified formal education as an important predictor of behavior (Prokopy et al., 2019). In addition, a number of authors have highlighted the positive relationship between the level of agricultural training and the adoption of conservation agriculture (Bielders et al., 2003; Rodríguez-Entrena & Arriaza, 2013; Serebrennikov et al., 2020).

An important factor toward climate-just behavior is *self-efficacy* – first introduced by Albert Bandura (Bandura, 1977; see also Gifford & Nilsson, 2014). Self-efficacy refers to an individual's conviction in their capacity to exert control over their own functioning and the events that impact their lives. An individual's perception of their own self-efficacy can serve as the basis for their motivation, well-being and personal achievements (Bandura, 1977). People's beliefs in their efficacy are developed by influences such as mastery experiences which encompass the instances in which individuals undertake new challenges and achieve accomplishment or social models which serve as a significant source of self-efficacy through the provision of vicarious experiences. According to Bandura (1977), observing individuals who are similar to oneself accomplish through persistent effort increases the observers' belief in their own ability to master similar tasks and achieve success. Furthermore, receiving affirmative verbal feedback while engaging in an intricate job convinces an individual to believe in their competence and ability to achieve success. To sum up, people who feel they have the self-efficacy to carry out a certain behavior are more likely to do so than those who perceive themselves as lacking the ability to behave in the desired way. The notion of self-efficacy may also be a matter of consumer sovereignty and environmental responsibility (Ericson et al., 2014; Fischer & Barth, 2014; Girod et al., 2014; Peattie, 2010). Self-efficacy is related to several other factors of the integrative model such as motivation, emotions, social norms and moral development.

Personality traits also influence climate-just behavior, with the most commonly established being the "Big Five: openness to experience, conscientiousness, extraversion, agreeableness and neuroticism" (McCrae & Costa, 1999). Kaiser and Byrka (2011) emphasize that with a more trait-like measurement, "people's environmental engagement can be predicted with up to 80–90%" (p. 72), in particular reflecting prosocial dispositions in relation to pro-environmental behavior. In two different studies, greater environmental concern was related to greater openness and agreeableness (the tendency to be compassionate and cooperative rather than suspicious and antagonistic toward others) (Hirsh, 2010; Klein et al., 2019). Another study has proven the significance of openness and conscientiousness in farmers' decisions (Sok et al., 2016). In a wide-ranging set of studies, openness, agreeableness and conscientiousness were strongly linked to environmental engagement across both individuals and nations (Milfont & Sibley, 2012). Conscientiousness is an important personality trait because a lack of it leads to impulsive behavior and a loss of self-control in the face of tempting unsustainable situations and is also related to habits which are often unconscious (see below under the "Habits" section; Ericson et al., 2014).

Values, Attitudes and Moral Development

Values are important to consider for achieving climate justice because they are the "guiding principles" and "desirable transsituational goals" that individuals use to judge situations: a person's sense of right and wrong or what "ought" to be (Darnton & Evans, 2013). Attitudes can be defined as a disposition to react favorably or unfavorably to an object, person, institution or policy. They are evaluative, that is, directed at a given object or target, for example, that landscapes are to be cultivated and shaped. Although attitudes are assumed to be relatively stable, evaluations can change rapidly as events unfold and new information becomes available (Ajzen, 2005; Schwerdtner Máñez et al., 2024).

In general, values are more abstract concepts than norms and attitudes, which usually refer to specific actions, objects and situations. In addition, "people's values form an ordered systems of value priorities that characterize them as individuals" (Schwartz, 2006, p. 1). In the Schwartz Value Survey opposing values are benevolence ("preserving and enhancing the welfare of those with whom one is in frequent personal contact") with achievement ("personal success through demonstrating competence according to social standards") and universalism ("protection for the welfare of all people and for nature") with power ("social status and prestige, control or dominance over people and resources") (Schwartz, 2006, p. 11). Clayton (2018) found that political orientation is determining attitudes toward environmental policy with an overall dislike of environmental policies being characteristic of conservatives.

Gould et al. (2023) have analyzed how 134 theories of human behavior treat values, which they define broadly to include value(s) related to both principles (e.g. moral values) and value(s) related to importance (e.g. cost or priorities). They found that "values and closely related constructs comprise roughly a third of all constructs ($n = 2232$) in analyzed theories" (p. 101355). The nuanced portrayal of values–behavior links offered by Gould et al. is crucial for understanding how values may be associated with transformative change.

Undoubtedly, conflicting values and perspectives on the climate crisis can arise due to individuals holding multiple and occasionally contradictory values (see above; Lindenberg & Steg, 2007; Howell, 2013; Howes & Gifford, 2008; Stern, 2000). In general, individuals with cooperative (prosocial) orientations emphasize joint gains between self and others, whereas those with competitive and individualistic orientations (pro-self) emphasize gains to themselves (Howell, 2013; Kaiser & Byrka, 2011; Lindenberg & Steg, 2007; Reese & Jacob, 2015; Steg et al., 2014). Individuals who are more people-oriented and less authoritarian (Schultz & Stone, 1994) have higher levels of moral development (Swearingen, 1990 and see below); and believe their actions will make a difference tend to be more environmentally concerned (Antonetti & Maklan, 2014; Howell, 2013). However, a barrier to low-carbon behavior originates from inadequate self-centeredness. It seems useful to take a more detailed look at possible reasons for it, such as a lack of *moral development*.

Understanding why moral reasoning – the cornerstone of ethical behavior – is not always possible is significantly supported by Kohlberg's Theory of Moral Development (1971, 1983). Regarding the explanation for low-carbon behavior, a number of recent studies make the case for taking Kohlberg's theory into account (Pinquart & Silbereisen, 2007; Stengel, 2011). According to Kohlberg's findings, moral development is primarily concerned with fairness and happens throughout an individual's lifetime (Kohlberg, 1983). There are three phases of moral development: preconventional, conventional and post-conventional (Kohlberg, 1971). The main premise of Kohlberg's scale, which examines how people justify their actions, is that moral behavior is more accountable, dependable and consistent in those who possess higher moral standards. Numerous empirical studies unequivocally demonstrate the universal applicability of Kohlberg's theory, in a wide range of cultural contexts (e.g. in The Bahamas, Mexico, Puerto Rico, Honduras, India, Pakistan, Indonesia,

Israel, Turkey, Iran, Taiwan, Thailand, Japan, New Zealand, Nigeria, Iceland, the United Kingdom, Finland, Germany and Poland) (Stengel, 2011).

Preconventional moral reasoning primarily considers repercussions that occur outside of the self and is egocentrically focused only on the self and is typified by an obedience-and-punishment orientation and/or by a self-interest orientation. It may be argued that far too many individuals continue to operate at this level, which is a significant barrier to progress toward lessening denial and eventually stepping up climate action. Adopting low-carbon behavior is only motivated by egoism, such as the desire to avoid penalty or take advantage of expected benefits, such as financial gain. Conventional morality is defined by people placing a high value on other people's approval and upholding positive interpersonal relationships through social norms and conformity and/or is marked by authority and upholding social order orientation. At this point, when morality is still primarily determined by external influences, the majority of active members of society still exist (Kohlberg, 1983). People internalize authority without challenging it and they base their decisions on the standards of the group they are a part of. At this stage, a person abides by social norms and rules even in the absence of penalties for disobedience or compliance. However, norms and customs are followed somewhat rigidly and the propriety or justice of a rule is rarely questioned (Kohlberg, 1983). It is possible to identify some chances here to encourage people to adopt less damaging behavior, but this will only happen if social norms become less consumption-oriented and encourage better low-carbon behavior.

At the postconventional level individuals follow their own set of moral principles, which encompass fundamental human rights like life, liberty and justice and are applicable to everyone on the planet rather than just members of a specific group (such as family, community or fellow citizens). People behave locally while adhering to the idea of global contexts in their reasoning. For them, laws are not unquestioning commands given by a political structure or other authority that are to be followed without inquiry. Furthermore, the person is willing to take action to uphold these values, even if doing so involves defying society's norms and suffering the repercussions of rejection (Stengel, 2011).

Individuals who arrive here should be considered the great hope for the transition to low-carbon behavior because, as Kohlberg highlights, the likelihood of converting moral judgments into appropriate action rises with the degree of moral development attained. At this stage, there is a minimum 75% correlation between judgment and action. Additionally, people are willing to accept a wide range of varied "costs" to put their moral judgment into practice as well as to overcome a lot more and higher situational hurdles. This is the case because at each higher moral level, there is a reduction in behavior patterns that might be used as an excuse, such as moral disengagement, due to the increasing consistency between thought and behavior (Kohlberg, 1996; Stengel, 2011).

Since conventional people derive their moral beliefs from the people around them and very few of them consider ethical concepts for themselves, people on the postconventional level may ideally act as role models for those on the conventional level (Kohlberg, 1983). If postconventional individuals would receive more attention they could act as pioneers of the required moral and behavioral revolution. While moral behavior can be influenced by outside factors, such as educational institutions, it is especially important to perceive real-world examples and role models of morally righteous (in this case climate-just) behavior because it reduces the costs associated with changing one's own lifestyle.

Motivation, Emotions and Psychological Needs

The basic drivers underlying human behaviors appear simple: namely "to seek out and attain rewards and to avoid punishments or penalties" (Blaukopf & DiGirolamo, 2007, p. 626). However, it becomes clear that the process is more complex than on focusing on the different existing rewards: the internal factors include basic needs for survival and reproduction (primary rewards) but can be more abstract and cognitive in nature (secondary rewards). Cultural values that must be learned, such as thinking monetarily, acclaim, security, knowledge and praise are also associated. Of course, rewards that produce feelings of pleasure and liking reinforce the behavior that achieves them (Blaukopf & DiGirolamo, 2007). The anticipation of rewards leads to motivated behavior that can also be labeled goal-directed behavior (Blaukopf & DiGirolamo, 2007). The three main motivations for behavior include (1) hedonic ones, which lead individuals to seek ways to improve how they feel; (2) those that sensitize individuals to gains or losses in changes in their financial or other resources; and (3) normative ones, which are concerned with the correctness of their behavior (Howes & Gifford, 2008; Lindenberg & Steg, 2007; Steg et al., 2014).

A similar and overlapping explanation for human motivation is the psychological needs meta-theory introduced by Eyster et al. (2022) (see above) which regards subjective well-being resulting from the satisfaction of the following six psychological needs as the ultimate purpose of human action: (1) pleasure promotion (the need to explore and approach enjoyable experiences, to self-actualize and seek out and understand novel arenas); (2) pain prevention (the need to manage and avoid painful experiences); (3) consonance (the need for consistency, including with values; for the world to make sense; and for stable self-identity); (4) competence (the need for efficacy and an important feature of "flow" activities; Csikszentmihalyi, 1990); (5) relatedness (the need to belong to secure relationships) and (6) autonomy (the need for ownership over one's actions, i.e. internal perceived locus of causality; Eyster et al., 2022, p. 738). The last three needs are also the core of Self-Determination Theory by Ryan and Deci (2000) which can help to understand human motivation and personality much better because it explains "people's inherent growth tendencies and innate psychological needs that are the basis for their self-motivation and personality integration, as well as for the conditions that foster those positive processes" (p. 68). It is necessary to consider psychological needs when the right incentives for approaches toward less carbon behavior have to be selected. The thoughts and studies of Ryan and Deci (2000) on motivation in general and intrinsic and extrinsic motivation in particular are important insights to explain human behavior and its change, also toward low-carbon behavior (see Chapter 5).

Closely related to psychological needs and motivation in general are emotions: emotional involvement – understood as the ability to experience an emotional reaction when experiencing the multifaceted impacts of the human-made climate crisis – holds strong importance for achieving climate justice (Jefferson et al., 2015; Stoll-Kleemann, 2019; Stoll-Kleemann et al., 2022). Kollmuss and Agyeman (2002) conclude, "the stronger a person's emotional reaction, the more likely that person will engage in a new behavior" (p. 254). Emotions also have a crucial role in making moral decisions and comprehending the moral consequences associated with the dangers of the climate crisis (Landmann, 2020). Roeser (2012) contends that emotions could serve as the crucial element in effectively conveying information about the climate crisis. Climate change emotions are consistently identified as one of the most powerful factors in predicting climate change risk perceptions, engagement in actions to mitigate climate change, adaptation behavior, support for policies and acceptance of technology (Brosch, 2021). In particular, three types of emotions are of particular relevance in this context, namely self-condemning emotions such as guilt, self-praising emotions like pride and other-suffering emotions such as empathy.

For example, Hurst and Sintov (2022) examine multiple studies that demonstrate the potential positive impact of pride and guilt on pro-environmental behavior. They discovered that the impact of these two emotions is contingent upon the specific circumstances in which they are elicited. In summary, their research findings offer continuous evidence that guilt plays a more significant role in encouraging behavior change (Hurst & Sintov, 2022). Additionally, they show that creating feelings of pride can be effective in some situations. The authors claim that there is a tendency toward negativity bias, leading to a greater and more consistent influence of guilt compared to pride. Hurst and Sintov (2022) propose that a negativity bias suggests that bad events signal the necessity for change, while positive events suggest that there is no need to alter behavior since things are going well. We adhere to this understanding because negative emotions are intensely unpleasant and hence can serve as a strong incentive for behavior aimed at altering the circumstances (Stoll-Kleemann et al., 2022).

It is necessary to repeatedly confront and emotionally engage those individuals who have significant carbon footprints if we make the assumption that individuals have a desire to avoid unfair circumstances, even when they personally gain from them (Kals et al., 2001; Montada et al., 1986; Sabbagh & Schmitt, 2016). This confrontation should effectively illustrate how their excessive emissions contribute to climate injustice. According to the paradigm of affect generalization, repetition plays a crucial role as a single emotional experience may not be adequate to induce behavioral change. According to Landmann (2020), emotions have an impact on behavioral intentions only if they are generalized to attitudes. The media can play a part in this recurrence. Various forms of media, including television, images, videos and newspapers, have been effectively utilized in emotion-based psychology studies to elicit specific feelings (see below and Chapter 5; Franikowski et al., 2021; Junge & Reisenzein, 2013).

Finally, anger is one of the most dominant emotions people feel in relation to the climate crisis (Landmann & Rohmann, 2020). A recent sample of more than 2,000 participants in Germany, which was quota-based according to sociodemographic characteristics, showed that across different social milieus, anger (after helplessness and disappointment) is the third most common emotion toward the climate crisis (Krause & Gagné, 2021). According to this study, anger is even slightly more common than fear when people are dealing with the climate crisis. This is confirmed by an Australian study in which people indicated that their most common emotional reaction to climate change was frustration (an aspect of climate anger) (Stanley et al., 2021). In the study with the German sample (Krause & Gagné, 2021), not only emotions but also justice issues were surveyed. This allowed the authors to show that the climate policy of the German government (as of before the Bundestag elections in September 2021) is predominantly assessed as unjust (64%) instead of just (36%). Participants also indicated that, in their perception, the public debate on climate change disproportionately considers the interests of richer rather than poorer people. Furthermore, the fact that perceptions of injustice in the climate context are associated with anger has been empirically demonstrated several times (Landmann & Rohmann, 2020; Reese & Jacob, 2015; Van Zomeren et al., 2008). The more people are angry about global, environmental and intergenerational injustice, the more likely they are to want to engage in environmental protection (Reese & Jacob, 2015).

Habits and Fast Thinking (Heuristics and Biases)

Habits are a barrier to climate-just behavior because they are highly unconscious and embedded in routines and social practices (Heimlich & Ardoin, 2008; Southerton, 2013; Verplanken & Roy, 2016). For example, many of our day-to-day carbon behavior as eating, mobility or energy-related decisions are routine, in that we act, decide and buy often and without much deliberation, e.g. leave the light on, take the car, buy unsustainable products. These habits are difficult to change, as they are rewarding because they save time and energy (via the routines) (Lewin, 1951; O'Riordan & Stoll-Kleemann, 2015; Stern, 2000).

The trend toward "comfort" has been a major influence on carbon-related habits encouraged by a lack of time and skills. This demonstrates how behaviors are facilitated by the structures of the production and supply systems. The so-called consumption-happiness myth – which is based on neuroscience research – explains how we are locked in consumption patterns based on mechanisms like habit formation, which interact to influence "our sense of self at the very deepest levels of consciousness" (Brannigan, 2011, p. 85).

A further recent study from neuroscience has shown that participants – when instructed to think about ways to increase their pro-environmental behaviors (e.g. taking the train) – showed increased activity in brain regions involved in reward integration. Conversely, when those participants were instructed to think about decreasing environmentally harmful behaviors (e.g. lowering the heating), they showed increased activity in regions involved in loss anticipation and cognitive control (Brevers et al., 2021). Increasing pro-environmental behaviors is more feasible than decreasing their environmentally harmful behaviors (Doell et al., 2021). Doell et al. (2023) emphasize that "this dissociation at the neural level may help to better understand why people are able to adopt new pro-environmental behaviors while simultaneously continuing to persist with environmentally harmful habits" (p. 1288). Further studies based on the considerations of gains versus losses and the integration of different types of value (e.g. economic versus social) also show that it is useful to integrate research from neuroscience and neuroeconomics (Doell et al., 2023; Sawe, 2017; Sawe & Chawla, 2021).

If it comes to "cognitive limits", it is necessary to look at the challenges the climate crisis poses to our perceptual, cognitive and affective information-processing system making it difficult to engage with. Here it is important to mention the concept of "fast thinking" which is associated with cognitive processes that involve quick, automatic and intuitive decision-making (in contrast to "slow thinking", which refers to more deliberate, conscious and analytical cognitive processes) (Kahneman, 2011). Fast thinking involves automatic, subconscious and intuitive mental processes and relies on heuristics and mental shortcuts, making it susceptible to biases and errors.

In addition to the loss aversion bias described above, there are much more cognitive biases that have a substantial influence on high-carbon activities by shaping individuals' perception and response to the climate crisis (e.g. Kahneman, 2011; Kahnemann et al., 1982; Gigerenzer et al., 2011; Thaler & Sunstein, 2022). The most important of them are (1) hyperbolic discounting because it refers to the tendency of individuals to emphasize present rewards over future advantages or repercussions; (2) the status quo bias because it characterizes the tendency to favor the existing state of affairs which is also a good explanation for the opposition toward modifications that would diminish carbon emissions; (3) the optimism bias which encompasses that numerous individuals hold the belief that they possess a lower probability than others of encountering adverse circumstances. This cognitive bias might result in the underestimation of one's own susceptibility to the adverse effects of the climate crisis; (4) the confirmation bias describes the tendency to selectively focus on information that aligns with preexisting beliefs, while disregarding information that contradicts them; and (5) the bystander effect refers to a situation in which individuals are unwilling to take action on climate change because they assume that others will handle it. This results in a diffusion of responsibility and a lack of effort in reducing individual carbon emissions. Comprehending and tackling these cognitive biases and heuristics is essential for devising efficient environmental legislation, educational campaigns and interventions to promote low-carbon behaviors. Possible solutions encompass streamlining sustainable options, efficiently conveying the immediate advantages of these options and transforming societal standards to establish low-carbon behavior as the default choice (e.g. see Kahneman, 2011; Kahnemann et al., 1982; Gigerenzer et al., 2011; Thaler & Sunstein, 2022).

External Factors

It is sometimes difficult to draw a sharp line between internal and external factors that influence climate-related behavior, and the interactions between the two levels are very strong. It is not only the case that external factors such as the political-economic system have an impact on individual behavior, but it can also be the other way around, as Lubchenco et al., 2016 highlight: "behaviors of individual actors at the local scale influence interactions at the regional or global scale" and "the collective effect of individual behaviors influences the larger-scale properties such that actors adapt to the changing conditions of the system context" (p. 14508). Furthermore, it is important to consider the aggregative nature of individual climate-related behavior, insofar as "even if an act harms no one, this act may be wrong because it is one of a set of acts that together harm other people" (Parfit, 1987; Peeters et al., 2015, p. 76).

Situational Context: Availability of and Access to (Low-Carbon) Infrastructure and Products

On the one hand, *infrastructure*, for example, with its transportation and energy systems are significant contributors to greenhouse gas emissions. Vehicle exhaust from roads, airports, ports and trains, as well as the combustion of fossil fuels in power plants and refineries, contribute to emissions of carbon dioxide (Kemfert, 2010). The process of constructing and upkeeping infrastructure necessitates substantial quantities of energy and resources, hence exacerbating emissions. On the other hand, for example, improved urban and land use planning can have significant positive impacts on low-carbon development and transportation systems. Urban areas that are dense and easily navigable, with efficient public transit networks, can mitigate the necessity for automobile commuting and decrease emissions resulting from transportation. This encompasses the availability of a good and affordable public transport system and a good cycle path network as well as the availability of sufficient energy and heat from renewable sources and well-insulated housing (Kemfert, 2010).

Overall, the availability of and access to low-carbon infrastructure and products is important for adopting climate-just behavior (Kemfert, 2010). This relates to all areas that are important for the transition to a low-energy society, not only mobility but also energy, food and consumption as a whole. For the food sector, this would include specific factors of food supply in the environment, such as the type of food, food sources and the availability of and access to food (Dagevos & Voordouw, 2013; Furst et al., 1996; Stoll-Kleemann & Schmidt, 2017; Verain et al., 2015).

Situational Context: Media Influence

As mentioned above in the section "emotions", people acquire comprehension and experience emotional reactions when watching media information (Moser & Dilling, 2011). Therefore, it is important to examine how *media influences* the cognitive, emotional and behavioral aspects that impact carbon-related behavior. Specifically, climate change communication through the media should make the link between one's own advantages and high carbon harmful actions, as well as the disadvantages faced by others, more evident. As explained above, it is crucial to have suitable and pertinent media coverage because these detrimental repercussions of the climate catastrophe are abstract, temporally and spatially remote, complex and unintentional (Heald, 2017). Hence, the media ought to include coverage of detrimental outcomes and effective measures to alleviate and adjust to such circumstances. Thus far, it is evident that the extensive information regarding climate change in the media frequently does not effectively inspire behavioral changes in its viewers (Markowitz & Shariff, 2012).

An obstacle to inspiring low-carbon behavior through the media is the challenge of determining reliable sources of information and guidance. This uncertainty arises from the media's tendency to present a mixture of ideas and arguments (Markowitz & Shariff, 2012). Moreover, the assortment of contradictory information regarding the climate catastrophe disseminated through various media outlets has significantly contributed to a state of perplexity. In particular, the false balance mechanism plays a major role here. It occurs when journalists – in an attempt to demonstrate impartiality and fairness – treat all perspectives as equally valid or significant, irrespective of the evidence or credibility behind each viewpoint (Brüggemann & Engesser, 2017; Rahmstorf, 2012). This false balance leads to the inaccurate comparison between widely proven scientific evidence and unsubstantiated perspectives, resulting in the dissemination of incorrect information and a state of uncertainty among the public. This mechanism is exemplified by the act of portraying climate change as a contentious issue, wherein equal weight is given to the opinions of climate scientists and those who deny the existence of climate change (Brüggemann & Engesser, 2017; Rahmstorf, 2012). The false balance can impede policy and action, particularly in the context of climate change. It can result in a lack of action or delayed action due to the false appearance of an ongoing debate or uncertainty, despite the absence of any such debate or doubt.

Hence, it is crucial to have communications that accurately represent the scientific consensus disseminated across various media platforms. Although there may be differing opinions on certain aspects among the scientific community, the fundamental arguments supporting manmade climate change are becoming more widely acknowledged (Stoll-Kleemann et al., 2022). Goldberg et al. (2019) contend that the public's comprehension of this scientific consensus serves as a "gateway belief". Individuals who become aware of the prevailing consensus are more inclined to believe that climate change is occurring, caused by human activity and poses a significant threat. Consequently, they are more likely to endorse climate change policies. Hence, it is crucial to provide context when using the language of risk, as it may be unfamiliar to many people. This language is becoming more prevalent in climate change communication. It is important to explain how science operates and emphasize the overwhelming scientific consensus on human-induced climate change. One common tactic employed by persons who deny the existence of the climate catastrophe is to deliberately create confusion among the public regarding the scientific consensus. This is achieved with the intention of obstructing or postponing political actions aimed at addressing climate change (Dunlap & McCright, 2015; Painter, 2013; see also Chapter 4). One such factor contributing to the media's inability to inspire changes in behavior connected to carbon is the fluctuating levels of media coverage on the climate catastrophe (Markowitz & Shariff, 2012).

It is worth noting that global media coverage of the climate catastrophe has significantly decreased since the start of the COVID-19 pandemic (Stoll-Kleemann et al., 2022). In addition, we were intrigued by the subjective media portrayal of climate change in contrast to the COVID-19 epidemic which they investigated in a representative German quota sample with 1,100 participants (Stoll-Kleemann et al., 2022). The respondents' answers demonstrate the notable prevalence of COVID-19 reporting. Specifically, 51% of the participants observed a significantly higher amount of news coverage regarding COVID-19 compared to climate change. A majority of the study respondents were unaware of the significant mortality rate resulting from heatwaves in 2018, which amounted to 69%. Merely 15% demonstrated acquaintance with this material, but 16% had skepticism over its accuracy. Undoubtedly, making such comparisons is challenging since individuals may not accurately recall specific figures presented by the media and may also be hesitant to assess one form of distress in relation to another. Nevertheless, it can be assumed that had the media coverage of the deaths in 2018 been more prominent, a greater number of individuals would have retained the specifics. Referencing a simple numerical value (exceeding 10,000 fatalities due to heat in 2018) within the framework of climate change evoked an emotional response in nearly half of the participants (47%), whereas the other half remained unaffected by this data (Stoll-Kleemann et al., 2022).

Stoll-Kleemann et al. (2022) assume that individuals may fail to accurately assign the knowledge to either climate change or their own actions. The inherent abstractness of climate change, characterized as "nonintuitive and cognitively demanding to comprehend" (Markowitz & Shariff, 2012, p. 244), may serve as an additional explanation for the lack of emotional impact. Markowitz and Shariff (2012) assert that comprehending climate change as a moral obligation does not happen spontaneously, at an instinctive level. Instead, it necessitates the use of chilly, cognitively challenging and ultimately less inspiring moral reasoning. Furthermore, these persons may encounter moral disengagement, as explained in Chapter 4. As previously said, Stoll-Kleemann et al. (2022) proceeded to inquire about the feelings of the participants who were emotionally impacted by the facts of the number of deaths caused by the 2018 heatwave in Germany and requested them to provide more specific details about their emotions. The data comprise 269 reported answers from a total of 248 individuals. The prevailing emotional reaction was shock and surprise, accompanied by feelings of being moved and experiencing pain or grief. Infrequently, the individuals expressed emotions such as anger, fear/worry, sympathy, helplessness and, to a lesser extent, guilt (Stoll-Kleemann et al., 2022). Thus, moral motivation can be stimulated by exposing individuals to both, real-life occurrences and the circumstances of others which in turn may stimulate the inclination to re-evaluate one's own conduct and redirect attention toward safeguarding the susceptible from preventable hardships.

According to Stoll-Kleemann et al.'s (2022) findings, the participants are likely to modify their behavior linked to climate change in response to media coverage similar to that of the COVID-19 pandemic. Participants may have effectively associated their carbon behavior with climate change consequences for others. Furthermore, it is possible that they had foreseen that, upon being informed of the repercussions of the climate issue, they would decrease their carbon emissions. This corroborates the argument regarding result attribution. These findings align with a study conducted by Holbert et al. (2003), which demonstrated that individuals who are concerned about the environment are more likely to watch television news and nature documentaries and this media consumption contributes to pro-environmental behaviors. Additionally, a study conducted in Taiwan by Huang (2016) clearly shows that media coverage of global warming has an impact on people's carbon-related behavior (Holbert et al., 2003).

Unfortunately, what the study suggests, namely that governments and organizations could utilize the media as a means of promoting environmental actions by individuals and actively market policies and efforts to mitigate climate change through various media channels is not (or only to a very limited degree) happening. Instead, we observe the insufficient media coverage explained above. This can be attributed to the application of Herman's and Chomsky's Propaganda Model (Herman & Chomsky, 2010; Stoll-Kleemann et al., 2022). This model provides a framework for comprehending the operations of the mainstream media and its associations with government propaganda requirements. Herman and Chomsky contend that the media function as instruments of influential stakeholders who exercise control and provide financial support for their activities. This is achieved by the careful selection of staff who possess the appropriate mindset, as well as by ensuring that editors and working journalists internalize the institution's strategy and adhere to its priorities and definitions of what is considered newsworthy. Herman and Chomsky (2010) aim to establish several media platforms, such as those operated by grassroots movements and nonprofit organizations, to more accurately represent the viewpoints of regular individuals and thereby promote the democratization of information dissemination.

Sociocultural, Sociodemographic and Further Group Factors Such as Social Norms and Social Identity

Sociocultural aspects refer to the influences and conditions that arise from the interaction between society and culture. Culture, religion and the development of social identities and social norms are significant sociocultural variables that strongly influence people's perceptions toward climate justice and related behavior (Stoll-Kleemann, 2019).

The presence of individuals at events where decisions are made regarding actions that impact the climate can exert a significant influence on behavior. This aspect of social norms, which pertains to how individuals perceive the behavior that is considered normal by their socially connected peers, can function both as an obstacle and a chance. This is because people modify their behavior to control how they are perceived by others and create a specific impression on them (Higgs, 2015; also see Farrow et al., 2017; Griskevicius et al., 2010). Social norms can be conveyed explicitly through cultural practices within a specific context (Farrow et al., 2017; Higgs, 2015), and they are influenced by external influences. However, social norms can not only be a powerful force in both inhibiting and encouraging low-carbon behavior but are also important for group coherence and mediate between the identity of the individual and group (Darnton, 2008). Groups give a sense of social identity and a sense of belonging to the social world. The central hypothesis of the social identity theory is that members of an in-group ("us") will seek to find negative aspects of an outgroup ("them"), thus enhancing their self-image (McLeod, 2023). If confronted with novel, unknown or potentially threatening phenomena, existing ideas and practices are used as collective coping mechanisms (Buijs et al., 2012). A careful analysis of the respective social group is required to understand and potentially influence low-carbon behavior. While group behavior is driven by descriptive norms, which ensure conformity with the behavior of other group members, injunctive norms reflect the expectations of society (Schwerdtner Máñez et al., 2024).

Culture as such is a more indirect factor as it influences public perception, policy responses, individual and collective climate-related behavior. Nevertheless, understanding the cultural dimensions of the climate crisis is crucial for developing effective strategies to address it. Different cultures have varying levels of awareness and concern about the climate crisis. Cultural narratives, beliefs and values shape how communities perceive the urgency of the climate crisis and their responsibility toward addressing it (Giddens, 2006; Norgaard, 2011). Cultural norms heavily influence consumption habits, energy use, dietary preferences and transportation choices, all of which generate high amounts of greenhouse gas emissions (examples see below). Cultures that prioritize consumerism and material wealth have a larger carbon footprint while other cultures have a deep-rooted respect for nature and emphasize sustainable and low-carbon lifestyles. For example, cultures with strong traditions of cooperation and collective resource management are more successful in implementing low-carbon practices, for example, indigenous communities across the globe which rely on renewable resources and traditional ecological knowledge (Jackson, 2016). In addition, cultural factors influence political ideologies and economic systems, which in turn shape climate policies and actions (Giddens, 2006). For example, a culture that highly values individual freedom might resist regulations on carbon emissions, whereas cultures with a communal ethos might be more supportive toward mitigation options. Positive examples are the Scandinavian countries whose cultures climate responsibility plays a major role and with governments having implemented policies that promote renewable energy, sustainable urban planning and low-carbon transportation options like cycling and public transit (Jackson, 2016; Sachs et al., 2021). A very positive example is certainly Bhutan which is known for its philosophy of Gross National Happiness, which includes sustainable development and nature conservation as a core element and being not only carbon neutral but even carbon negative, thanks to its extensive forest cover and environmental policies (Jackson, 2016).

Overall, it is essential to recognize the heterogeneity in society's understanding of carbon-related behavior such as consumption, eating, heating or traveling with influencing variables including age, gender, social values or living in urban or rural areas or close to the sea. By understanding public perceptions of climate justice, it is, for example, easier to tailor respective campaigns. In the context of sociocultural factors, it could happen that a group could have similar knowledge of the climate crisis but respond to different engagement approaches differently. In the field of sustainable ocean behavior and literacy, Santoro et al. (2017) state that this varies among countries and cultures. In Europe, which is home to several basins and regional seas, it is important to take into account the diverse cultural settings and fully comprehend how individuals interact with the sea. One example from Bien Unido

Double Barrier Reef Marine Park in the Philippines where the practice of blast and cyanide fishing posed a threat demonstrates the possible influence of cultural and religious affiliations because finally submerging religious sculptures in water led to significant reductions in unlawful activities, therefore promoting the marine conservation goals of the Park by aligning these priorities with cultural values (Jefferson et al., 2015).

The importance of culture, religion, social norms and social demographic group factors for carbon-related behavior can also be demonstrated by the example of meat consumption: Fiddes (1992) proposes that the intake of red meat in Western culture is motivated by the aspiration to assert human dominance over the natural environment as a means of expressing power. Conversely, numerous cultures and faiths have intricate systems of taboos and restrictions in terms of consuming different kinds of meat (Beardsworth & Bryman, 2004). Haverstock and Forgays (2012) validate that individuals who transition to consuming animal products perceive these dietary patterns as integral to their cultural and religious heritage. Consequently, their food choices or abstentions may be influenced by the ahimsa principle, which advocates for nonharming of living beings and is a fundamental belief in religions such as Buddhism, Hinduism and Jainism. Joy (2011) affirms that most individuals who consume meat perceive it as a cultural convention rather than a deliberate decision, as it is not essential for their survival. People generally do not contemplate the reasons behind their repulsion toward consuming dogs while finding the consumption of cows appealing, or vice versa. They do not question the rationale behind consuming any type of animal (Joy, 2011, see also Piazza et al., 2015; Rauschmayer & Omann, 2012). Furthermore, Higgs (2015) elucidates that individuals adhere to such eating norms since it fosters a sense of belonging and popularity within a social group (Higgs, 2015). Finally, it remains a belief of a majority of men in the Western World that eating meat enhances the physical strength and vitality of males, which is a significant factor that diminishes the attractiveness of vegetarianism. This is due to the expression and fulfillment of stereotypes related to masculinity (Ruby & Heine, 2011; Vartanian, 2015). Schösler et al. (2015) determined that the presence of traditional perceptions of masculinity hinders the adoption of a diet that is less reliant on meat.

Schwerdtner Máñez et al. (2024) have summarized research that demonstrates how further sociodemographic variables are important for pro-environmental behavior within the group of farmers:

> A meta-analysis of 125 articles by Mozzato et al. (2018) finds that female farmers have a higher motivation to adopt environmentally friendly farming practices, especially in Europe and North America and that younger farmers prevail as early adopters for these practices, while older farmers act as followers. In his review, Burton (2014) concludes that demographic characteristics are influenced to varying extents by cultural–historical patterns leading to cohort effects or socialized differences. This is also

confirmed by Mozzato et al. (2018), who highlight the influence of the geographical context as well as temporal trends on both sociodemographic as well as other factors.

(p. 5)

Politico-Economic Factors

To achieve climate-just behavior ideally supportive government policies and practices would be exercised as well as new and different business practices and civil society initiatives would be working in synergy. Many political agreements exist on paper at various local, regional and global levels such as the UN Sustainable Development Goal 13 "Climate Action", the European Green New Deal and of course The Paris Climate Agreement. Unfortunately, the reality is that in many cases, major implementation problems are obvious based on a lack of political will, lobbyism and problematic power structures within an economic system based on unleashed capitalism. Certainly, the most depressing example in terms of political failure concerning the implementation of important agreements is the Paris Agreement, which aims to limit global temperatures "well below" 2°C above pre-industrial levels, with the ultimate objective of reducing this to 1.5°C (UNFCCC, 2015). We constantly observe with every new Conference of the Parties of the UN Climate Conferences that the probability of collective failure to achieve these goals is extremely high (Bode, 2018; Ekardt et al., 2018; Klein, 2014). In order to achieve the 1.5°C path, for example, coal extraction and burning and, of course, the burning of fossil fuels in general would have to be massively restricted immediately. Germany – the world's fourth-largest economy – is a concrete example of climate policy failure: its absolute coal use has increased in recent years by 11% (2009–2014), and it continues to provide significant subsidies to coal and has introduced new subsidies for coal-fired power some years ago (Bode, 2018; Climate Transparency, 2017; Whitley et al., 2017). The country will not stop burning coal until 2038. In addition, the GHG emissions of Germany's transport sector are also growing. This failure is caused by "lock-in" effects from existing and currently constructed energy and transport infrastructure and the influence of lobbyism (Bode, 2018; Balser & Ritzer, 2018; Klein, 2014; Stoll-Kleemann, 2019; Tillack, 2015).

Our current market-driven capitalistic system, which relies heavily on growth to promote consumption in all aspects of life, resulting in excessive consumerism, is a significant contributor to surpassing planetary limits (Felber, 2015; Jackson, 2016; Kemfert, 2010; Klein, 2014; Piketty, 2022; Saito, 2024). According to Amel et al. (2009), overconsumption occurs when people excessively desire to acquire things because they feel empty or that their life lacks meaning. They use material consumption as a way to alleviate these negative emotions. The issue of aggressive advertising that aims to create demand for "unnecessary" products is closely connected to the problem of planned obsolescence, where products are intentionally designed with a limited lifespan.

This strategy increases long-term sales by shortening the time between repeat purchases (Jackson, 2016). Additionally, it is worth noting that unsustainable options continue to be the default choice.

Not only consumerism but also other core principles of capitalism such as profit maximization, externalization of environmental costs, short-termism and systemic inequality contribute heavily to the climate crisis (Dolderer et al., 2021; Jackson, 2016; Piketty, 2022; Saito, 2024). For example, capitalism allows businesses to externalize environmental costs. This means that the environmental damage caused by production – such as greenhouse gas emissions and pollution – is not fully accounted for in the cost of goods and services. Instead, these costs are mainly borne by the public in the form of health problems, climate change impacts and environmental degradation (Jackson, 2016). As a result, companies do not have strong incentives to adopt more low-carbon practices. The same is true for the so-called race to the bottom principle in environmental standards. Countries might resist implementing strong environmental regulations because they assume losing competitive advantage or driving away investment.

In general, financial factors significantly impact individuals' consumption decisions, for example on meat consumption (Edjabou & Smed, 2013; Furst et al., 1996; Lanfranco & Rava, 2014; Ritson & Petrovici, 2001). When meat is inexpensive and other food is more expensive, this becomes a significant obstacle for individuals who want to reduce their meat consumption (Stoll-Kleemann & Schmidt, 2017). Conversely, in certain nations, meat is among the costliest food items in individuals' shopping carts. Consequently, reducing meat consumption enables individuals to economize and potentially upgrade to higher quality meat (Dibb & Fitzpatrick, 2014). Recent research shows that an internalization of external costs into the market price of agricultural products (true costs) closes the gap between organic and conventional products. Including the positive or negative impacts of a produced commodity into their production costs at least could partly correct current market distortions (see Chapter 5 and Michalke et al., 2022).

Box 1:   Digression on Inequality

Digression on Inequality (and its overall meaning for the climate crisis)

It is obvious that the behavior of powerful economic actors within the capitalistic system has led to a very unequal distribution of wealth and power which has exacerbated the climate crisis (see Chapters 1 and 2; Jackson, 2016; Klein, 2014; Saito, 2024). We have already demonstrated above that those richer individuals and nations, who consume and pollute much more have a very disproportionate impact on the climate and environment in general, while the poorest, who contribute least to the climate crisis, often suffer the most from its effects. Therefore, inequality is one major problem of the climate crisis, and it is interesting to know more about the features of it because eliminating or downsizing inequality would help decreasing greenhouse gas emissions. The adherence to the capitalist system by those who profit from it contributes to further already strong systemic inequality (Chomsky, 2004; Piketty, 2022) through various mechanisms and dynamics inherent in the system out of which five important ones are summarized here.

The first principle is the accumulation of wealth over time. Those who already possess wealth overly use their opportunities to invest, generate returns and accumulate even more wealth. This has created a cycle where wealth has generated more wealth, contributing to the concentration of resources in the hands of a few (Chomsky, 2004). Second, capitalism strongly supports the intergenerational transfer of wealth. Families with significant financial resources pass on their wealth to future generations, further entrenching economic advantages within certain families or social groups. This results in the perpetuation of inequality across generations (Piketty, 2022). Third, in capitalist economies in the 21st century, we observe the concentration of market power in the hands of a few dominant corporations such as Amazon, Google, Microsoft and Apple. These corporations have gained monopolistic or oligopolistic control, limiting competition and enabling them to set prices and influence markets to their advantage, contributing to economic inequality (Chomsky, 2004; Piketty, 2022). Fourth, in capitalist systems, individuals and businesses are driven by the pursuit of profit. This pursuit results in significant disparities in income between different individuals and groups. Executives and shareholders of successful businesses have accumulated substantial wealth, while lower wage workers struggle to meet their basic needs. The market-driven nature of capitalism constantly leads to the unequal distribution of income (Felber, 2015; Piketty, 2022; Stiglitz, 2013). Finally, in a capitalist system, access to opportunities such as education, healthcare and employment often depends on an individual's financial resources. Those with greater financial means have better access to quality education, healthcare services and networking opportunities, providing them with a competitive advantage in the workforce (Chomsky, 2004; Felber, 2015; Piketty, 2022; Stiglitz, 2013).

Efforts to influence politics through lobbying, which are often disguised in "creative" ways, have been "successful" in shaping laws and reducing regulations that originally aimed to protect the environment (see also Chapter 4). This has been extensively documented by various researchers (Bandura, 2007; Billé et al., 2013; Chomsky, 2004; Danciu, 2014; Klein, 2014; O'Riordan & Stoll-Kleemann, 2015; Stern, 2000). The fear of provoking criticism from influential interest groups is one factor contributing to the lack of political action in the past and present also regarding efforts to make agriculture more sustainable (Stoll-Kleemann & Schmidt, 2017; WBGU, 2018; Withana et al., 2012). Joy (2011) estimates that the animal agribusiness industry in the USA is worth $125 billion and is dominated by a small number of corporations. These corporations include agro-chemical and seed companies, processing companies, food manufacturers, food retailers, transportation systems, pharmaceutical companies and farm equipment producers. In other countries, such as Germany, the situation is nearly identical. In Sexton's (2013) study, it was discovered that power dynamics within the agricultural markets in the United States strongly support the consolidation of companies and the use of contract buying. As a result, the concept of free markets becomes obsolete in the food industry. An issue of major concern is the extent of subsidies on a global scale, which results in inefficiencies and misallocations within the market (Withana et al., 2012). The industrialized countries (OECD members) offer subsidies totaling $52 billion for livestock-based products, including animal feed and animal products which are particularly climate unfriendly (Heinrich-Böll Foundation, 2014). For example, meat in numerous nations is subjected to a lower value-added tax rate such as in Germany (Keller & Kretschmer, 2012; Stoll-Kleemann & Schmidt, 2017).

In a more recent study on climate lobbyism in the United States by Brulle (2018) has been demonstrated that "climate legislation has often been the focus of intense lobbying" and that "despite the introduction of several major bills to limit carbon emissions in the USA, none of them have been passed" (p. 289) mainly with lobbying as the most important influence. Brulle (2018) also provides a calculation that over the 16-year period from 2000 to 2016, more than $2 billion was expended on climate lobbying in the US Congress. The overwhelming majority of climate lobbying expenses originated from industries that would be significantly affected by climate legislation, including fossil fuel and transportation corporations, utilities and related trade associations. The expenditures made by these sectors greatly surpass those of environmental organizations and renewable energy firms. The levels of lobbying expenses seem to be associated with the introduction and likelihood of passing substantial climate legislation (Brulle, 2018).

These data further demonstrate the constraints of science advocacy endeavors. Environmental organizations' climate lobbying efforts account for a mere 3% of the overall lobbying expenditures. Hence, the substantial financial outlays and uninterrupted presence of professional lobbyists curtail the influence of volunteer climate advocates. Furthermore, lobbying is a hidden activity. There is a lack of public discussion or rebuttal of perspectives presented by professional lobbyists who have confidential meetings with government leaders. Consequently, the lobbying process has the power to greatly change the control that government decision-makers have over the type and flow of information. This leads to a situation where communication is consistently distorted (Chomsky, 2004). This procedure may impede the dissemination of precise scientific information into the decision-making process. Critics have strongly criticized lobbying, arguing that it poses a threat to democracy by establishing exclusive networks of influential decision-makers that marginalize the general population (Chomsky, 2004; Edwards, 2016 in Brulle, 2018).

At present, global, national and regional efforts to solve the climate crisis are alarmingly limited. In Chapter 5, we will discuss whether and how, against the backdrop of successful climate lobbying to prevent effective climate policy, the behavior of individuals, communities and pioneers from different sectors can still become a decisive means of influencing political and economic decision-makers, for example through consumer power and/or established democratic mechanisms, to take account of climate justice concerns (see e.g. Felber, 2015). What is clear thus far is that in order to achieve a more effective implementation of climate policies and international agreements, it is essential to improve capacities in terms of participation, climate communication and responsible consumption (abstention) decisions and promote a greater willingness of the public to participate, for example through municipal measures or engagement in civil society organizations regarding a fairer climate policy.

Overall, effective solutions for a climate-just society need to combine the knowledge we have provided in this chapter on personal factors such as motivation, emotions, moral development, habits and cognitive processes with external factors such as sociocultural and political-economic factors. The complexity of human behavior, embedded in a complex globalized world, requires well-thought-out interventions and approaches such as a combination of regulatory measures, market incentives, improved media coverage and changes in consumer behavior. It is questionable whether climate justice measures have a chance of being implemented in a capitalist framework as long as economic interests are so powerful and focused exclusively on material growth. Potentially radical and far-reaching large-scale changes needed to tackle the climate crisis – such as a significant shift away from fossil fuels – are only possible in a system where lobbying and inequality can be overcome, for example through new cultural narratives, role models, emotions and symbols conveyed through the media to shape public discourse on the climate crisis. Further proposals are discussed in Chapter 5.

References

Ajzen, I. (1991). The theory of planned behavior. *Organizational Behavior and Human Decision Processes, 50*(2), 179–211. https://doi.org/10.1016/0749-5978(91)90020-t

Ajzen, I. (2005). *Attitudes, personality and behaviour.* McGraw-Hill Education UK.

Amel, E. L., Manning, C. M., & Scott, B. A. (2009). Mindfulness and sustainable behavior: Pondering attention and awareness as means for increasing green behavior. *Ecopsychology, 1*(1), 14–25. https://doi.org/10.1089/eco.2008.0005

Antonetti, P., & Maklan, S. (2014). Feelings that make a difference: How guilt and pride convince consumers of the effectiveness of sustainable consumption choices. *Journal of Business Ethics, 124*(1), 117–134. https://doi.org/10.1007/s10551-013-1841-9

Balser, M., & Ritzer, U. (2018). *Lobbykratie: Wie die Wirtschaft sich Einfluss, Mehrheiten, Gesetze kauft.* Droemer Knaur.

Bandura, A. (1977). Self-efficacy: Toward a unifying theory of behavioral change. *Psychological Review, 84*(2), 191–215. https://doi.org/10.1037/0033-295x.84.2.191

Bandura, A. (2007). Impeding ecological sustainability through selective moral disengagement. *International Journal of Innovation and Sustainable Development, 2*(1), 8–35. https://doi.org/10.1504/ijisd.2007.016056

Barreto, M., Szóstek, A., Karapanos, E., Nunes, N., Pereira, L., & Quintal, F. (2014). Understanding families' motivations for sustainable behaviors. *Computers in Human Behavior, 40*, 6–15. https://doi.org/10.1016/j.chb.2014.07.042

Barr, S., Gilg, A. W., & Shaw, G. (2011). Citizens, consumers and sustainability: (Re)framing environmental practice in an age of climate change. *Global Environmental Change, 21*(4), 1224–1233. https://doi.org/10.1016/j.gloenvcha.2011.07.009

Beardsworth, A., & Bryman, A. (2004). Meat consumption and meat avoidance among young people. *British Food Journal, 106*(4), 313–327. https://doi.org/10.1108/00070700410529573

Bielders, C., Ramelot, C., & Persoons, É (2003). Farmer perception of runoff and erosion and extent of flooding in the silt-loam belt of the Belgian Walloon region. *Environmental Science & Policy, 6*(1), 85–93. https://doi.org/10.1016/s1462-9011(02)00117-x

Billé, R., Kelly, R. P., Biastoch, A., Harrould-Kolieb, E., Herr, D., Joos, F., Kroeker, K. J., Laffoley, D., Oschlies, A., & Gattuso, J. (2013). Taking action against ocean acidification: A review of management and policy options. *Environmental Management, 52*(4), 761–779. https://doi.org/10.1007/s00267-013-0132-7

Blaukopf, C. L., & DiGirolamo, G. J. (2007). Reward, context, and human behaviour. *The Scientific World Journal, 7*, 626–640. https://doi.org/10.1100/tsw.2007.122

Bode, T. (2018). *Die Diktatur der Konzerne: Wie globale Unternehmen uns schaden und die Demokratie zerstören.* With the assistance of Christian Brückner. Fischer Verlag.

Brannigan, F. (2011). Dismantling the consumption-happiness myth: A neuropsychological perspective on the mechanisms that lock us into unsustainable consumption. *Engaging the public with climate change* (pp. 84–99). Earthscan.

Brevers, D., Baeken, C., Maurage, P., Sescousse, G., Vögele, C., & Billieux, J. (2021). Brain mechanisms underlying prospective thinking of sustainable behaviours. *Nature Sustainability, 4*(5), 433–439. https://doi.org/10.1038/s41893-020-00658-3

Brosch, T. (2021). Affect and emotions as drivers of climate change perception and action: A review. *Current Opinion in Behavioral Sciences, 42*, 15–21. https://doi.org/10.1016/j.cobeha.2021.02.001

Brüggemann, M., & Engesser, S. (2017). Beyond false balance: How interpretive journalism shapes media coverage of climate change. *Global Environmental Change, 42,* 58–67. https://doi.org/10.1016/j.gloenvcha.2016.11.004

Brulle, R. J. (2018). The climate lobby: A sectoral analysis of lobbying spending on climate change in the USA, 2000 to 2016. *Climatic Change, 149*(3–4), 289–303. https://doi.org/10.1007/s10584-018-2241-z

Buijs, A., Hovardas, T., Figari, H., Castro, P., Devine-Wright, P., Fischer, A., Mouro, C., & Selge, S. (2012). Understanding people's ideas on natural resource management: Research on social representations of nature. *Society & Natural Resources, 25*(11), 1167–1181. https://doi.org/10.1080/08941920.2012.670369

Burton, R. J. (2014). The influence of farmer demographic characteristics on environmental behaviour: A review. *Journal of Environmental Management, 135,* 19–26. https://doi.org/10.1016/j.jenvman.2013.12.005

Chaiken, S. (1980). Heuristic versus systematic information processing and the use of source versus message cues in persuasion. *Journal of Personality and Social Psychology, 39*(5), 752. https://fbaum.unc.edu/teaching/articles/jpsp-1980-Chaiken.pdf

Chomsky, N. (2004). *Profit over people: Neoliberalismus und globale Weltordnung.* Europa Verlag.

Cialdini, R. B., Demaine, L. J., Sagarin, B. J., Barrett, D. W., Rhoads, K., & Winter, P. L. (2006). Managing social norms for persuasive impact. *Social Influence, 1*(1), 3–15. https://doi.org/10.1080/15534510500181459

Clayton, S. (2018). The role of perceived justice, political ideology, and individual or collective framing in support for environmental policies. *Social Justice Research, 31*(3), 219–237. https://doi.org/10.1007/s11211-018-0303-z

Climate Transparency. (2017). *Brown to green: The G20 transition to a low carbon economy.* Climate Transparency, c/o Humboldt-Viadrina Governance Platform. https://www.climate-transparency.org/g20-climate-performance/g20report2017

Csikszentmihalyi, M. (1990). *Flow: The psychology of optimal experience.* Harper & Row.

Dagevos, H., & Voordouw, J. (2013). Sustainability and meat consumption: Is reduction realistic? *Sustainability: Science, Practice and Policy, 9*(2), 60–69. https://doi.org/10.1080/15487733.2013.11908115

Danciu, V. (2014). Manipulative marketing: Persuasion and manipulation of the consumer through advertising. *Theoretical and Applied Economics, 2*(591), 19–34. https://doaj.org/article/a128835485c2498fad24213a4462f1e8

Darnton, A. (2008). GSR behaviour change knowledge review. *Reference report: An overview of behaviour change models and their uses.* GSR Government Social Research. https://assets.publishing.service.gov.uk/media/5a7f0d5940f0b6230268d23a/Behaviour_change_reference_report_tcm6-9697.pdf

Darnton, A., & Evans, D. (2013). *Influencing behaviours. A technical guide to the ISM tool.* The Scottish Government.

DeCharms, R. (1983). *Personal causation: The internal affective determinants of behavior.* Psychology Press.

Dibb, S., & Fitzpatrick, I. (2014). *Let's talk about meat: Changing dietary behaviour for the 21st century.* Eating Better.

Diekmann, A., & Preisendorfer, P. (2003). Green and greenback: The behavioral effects of environmental attitudes in low-cost and high-cost situations. *Rationality and Society, 15,* 441–472.

Doell, K. C., Berman, M. G., Bratman, G. N., Knutson, B., Kühn, S., Lamm, C., Pahl, S., Sawe, N., Van Bavel, J. J., White, M. P., & Brosch, T. (2023). Leveraging neuroscience for climate change research. *Nature Climate Change, 13*(12), 1288–1297. https://doi.org/10.1038/s41558-023-01857-4

Doell, K. C., Conte, B., & Brosch, T. (2021). Interindividual differences in environmentally relevant positive trait affect impacts sustainable behavior in everyday life. *Scientific Reports, 11*(1), 20423. https://doi.org/10.1038/s41598-021-99438-y

Dolderer, J., Felber, C., & Teitscheid, P. (2021). From neoclassical economics to common good economics. *Sustainability, 13*(4), 2093. https://doi.org/10.3390/su13042093

Dunlap, R. E., & McCright, A. M. (2015). Challenging climate change: The denial countermovement. *Climate change and society: Sociological perspectives* (pp. 300–332). Oxford University Press.

Edjabou, L. D., & Smed, S. (2013). The effect of using consumption taxes on foods to promote climate friendly diets – The case of Denmark. *Food Policy, 39*, 84–96. https://doi.org/10.1016/j.foodpol.2012.12.004

Edwards, L. (2016). The role of public relations in deliberative systems. *Journal of Communication, 66*(1), 60–81. https://doi.org/10.1111/jcom.12199

Ekardt, F., Wieding, J., & Zorn, A. (2018). Paris Agreement, precautionary principle and human rights: Zero emissions in two decades? *Sustainability, 10*, 2812. https://doi.org/10.3390/su10082812

Ericson, T., Kjønstad, B. G., & Barstad, A. (2014). Mindfulness and sustainability. *Ecological Economics, 104*, 73–79. https://doi.org/10.1016/j.ecolecon.2014.04.007

Evans, J. S. B. T. (2008). Dual-processing accounts of reasoning, judgment, and social cognition. *Annual Review of Psychology, 59*(1), 255–278. https://doi.org/10.1146/annurev.psych.59.103006.093629

Eyster, H. N., Satterfield, T., & Chan, K. M. A. (2022). Why people do what they do: An interdisciplinary synthesis of human action theories. *Annual Review of Environment and Resources, 47*(1), 725–751. https://doi.org/10.1146/annurev-environ-020422-125351

Farrow, K., Grolleau, G., & Ibanez, L. (2017). Social norms and pro-environmental behavior: A review of the evidence. *Ecological Economics, 140*, 1–13. https://doi.org/10.1016/j.ecolecon.2017.04.017

Felber, C. (2015). *Change everything: Creating an economy for the common good.* Zed Books.

Fiddes, N. (1992). *Meat: A natural symbol.* Routledge.

Fischer, D., & Barth, M. (2014). Key competencies for and beyond sustainable consumption: An educational contribution to the debate. *Gaia (Heidelberg), 23*(3), 193–200. https://doi.org/10.14512/gaia.23.s1.7

Franikowski, P., Kriegeskorte, L., & Reisenzein, R. (2021). Perceptual latencies of object recognition and affect measured with the rotating spot method: Chronometric evidence for semantic primacy. *Emotion, 21*(8), 1744–1759. https://doi.org/10.1037/emo0000991

Furst, T., Connors, M., Bisogni, C. A., Sobal, J., & Falk, L. W. (1996). Food choice: A conceptual model of the process. *Appetite, 26*(3), 247–266. https://doi.org/10.1006/appe.1996.0019

Giddens, A. (2006). *The politics of climate change.* Polity Press.

Gifford, R., & Nilsson, A. (2014). Personal and social factors that influence pro-environmental concern and behaviour: A review. *International Journal of Psychology, 49*, 141–157. https://doi.org/10.1002/ijop.12034

Gigerenzer, G., Hertwig, R., & Pachur, T. (2011). *Heuristics: The foundations of adaptive behavior.* Oxford University Press.

Girod, B., Van Vuuren, D., & Hertwich, E. G. (2014). Climate policy through changing consumption choices: Options and obstacles for reducing greenhouse gas emissions. *Global Environmental Change*, *25*, 5–15. https://doi.org/10.1016/j.gloenvcha.2014.01.004

Goldberg, M. H., Van der Linden, S., Maibach, E., & Leiserowitz, A. (2019). Discussing global warming leads to greater acceptance of climate science. *Proceedings of the National Academy of Sciences of the United States of America*, *116*(30), 14804–14805. https://doi.org/10.1073/pnas.1906589116

Gould et al. 2023: Values in Theories of Human Behavior. Current Opinion in Environmental Sustainability 64,101355 before Griskevicius, V., Tybur.

Griskevicius, V., Tybur, J. M., & Van Den Bergh, B. (2010). Going green to be seen: Status, reputation, and conspicuous conservation. *Journal of Personality and Social Psychology*, *98*(3), 392–404. https://doi.org/10.1037/a0017346

Haverstock, K., & Forgays, D. K. (2012). To eat or not to eat: A comparison of current and former animal product limiters. *Appetite*, *58*(3), 1030–1036. https://doi.org/10.1016/j.appet.2012.02.048

Heald, S. G. (2017). Climate silence, moral disengagement, and self-efficacy: How Albert Bandura's theories inform our climate-change predicament. *Environment: Science and Policy for Sustainable Development*, *59*(6), 4–15. https://doi.org/10.1080/00139157.2017.1374792

Heimlich, J. E., & Ardoin, N. M. (2008). Understanding behavior to understand behavior change: A literature review. *Environmental Education Research*, *14*(3), 215–237. https://doi.org/10.1080/13504620802148881

Heinrich-Böll Foundation. (2014). *Meat atlas. Facts and figures about the animals we eat*. Heinrich-Böll Foundation.

Herman, E. S., & Chomsky, N. (2010). *Manufacturing consent: The political economy of the mass media*. Digital.

Higgs, S. (2015). Social norms and their influence on eating behaviours. *Appetite*, *86*, 38–44. https://doi.org/10.1016/j.appet.2014.10.021

Hirsh, J. B. (2010). Personality and environmental concern. *Journal of Environmental Psychology*, *30*(2), 245–248. https://doi.org/10.1016/j.jenvp.2010.01.004

Holbert, R. L., Kwak, N., & Shah, D. V. (2003). Environmental concern, patterns of television viewing, and pro-environmental behaviors: Integrating models of media consumption and effects. *Journal of Broadcasting & Electronic Media*, *47*(2), 177–196. https://doi.org/10.1207/s15506878jobem4702_2

Howell, R. (2013). It's not (just) "the environment, stupid!" Values, motivations, and routes to engagement of people adopting lower-carbon lifestyles. *Global Environmental Change*, *23*(1), 281–290. https://doi.org/10.1016/j.gloenvcha.2012.10.015

Howes, Y., & Gifford, R. (2008). Stable or dynamic value importance? The interaction between value endorsement level and situational differences on decision-making in environmental issues. *Environment and Behavior*, *41*(4), 549–582. https://doi.org/10.1177/0013916508318146

Huang, H. (2016). Media use, environmental beliefs, self-efficacy, and pro-environmental behavior. *Journal of Business Research*, *69*(6), 2206–2212. https://doi.org/10.1016/j.jbusres.2015.12.031

Hurst, K., & Sintov, N. D. (2022). Guilt consistently motivates pro-environmental outcomes while pride depends on context. *Journal of Environmental Psychology*, *80*, 101776. https://doi.org/10.1016/j.jenvp.2022.101776

Jackson, T. (2016). *Prosperity without growth: Foundations for the economy of tomorrow*. Routledge.

Jefferson, R., McKinley, E., Capstick, S., Fletcher, S., Griffin, H., & Milanese, M. (2015). Understanding audiences: Making public perceptions research matter to marine conservation. *Ocean & Coastal Management, 115*, 61–70. https://doi.org/10.1016/j.ocecoaman.2015.06.014

Joy, M. (2011). *Why we love dogs, eat pigs and wear cows. An introduction to carnism.* Conari Press.

Junge, M., & Reisenzein, R. (2013). Indirect scaling methods for testing quantitative emotion theories. *Cognition & Emotion, 27*(7), 1247–1275. https://doi.org/10.1080/02699931.2013.782267

Kahneman, D. (2011). *Thinking, fast and slow.* Penguin Books Ltd.

Kahnemann, D., Slovic, P., & Tversky, A. (1982). *Judgment under uncertainty: Heuristics and biases.* Cambridge University Press.

Kaiser, F. G., & Byrka, K. (2011). Environmentalism as a trait: Gauging people's prosocial personality in terms of environmental engagement. *International Journal of Psychology, 46*(1), 71–79. https://doi.org/10.1080/00207594.2010.516830

Kals, E., Maes, J., & Becker, R. (2001). The overestimated impact of self-interest and the underestimated impact of justice motives. *Trames Journal of the Humanities and Social Sciences, 55*, 269–287. https://doi.org/10.3176/tr.2001.3.05

Keller, M., & Kretschmer, J. (2012). Instrumente im Sinne einer nachhaltigen, klimafreundlichen Fleischproduktion. *Eine Untersuchung im Auftrag von MISEREOR.* Bischöfliches Hilfswerk MISEREOR e.V.

Kemfert, C. (2010). Sind wirtschaftliches Wachstum und ein Klimaschutz, der die Erderwärmung bei 2° plus hält, vereinbar? *Magazin der Heinrich-Böll Foundation, 1.* https://www.boell.de/sites/default/files/BoellThema_1-10_V06_abReader7kommentierbar.pdf

Klein, N. (2014). *This changes everything: Capitalism vs. the climate.* Allen Lane.

Klein, S. A., Heck, D. W., Reese, G., & Hilbig, B. E. (2019). On the relationship between openness to experience, political orientation, and pro-environmental behavior. *Personality and Individual Differences, 138*, 344–348. https://doi.org/10.1016/j.paid.2018.10.017

Kohlberg, L. (1971). Stages of moral development as a basis for moral education. *Moral Education, 1*(51), 23–92. https://doi.org/10.3138/9781442656758-004

Kohlberg, L. (1983). *Essays on moral development. The psychology of moral development: The nature and validity of moral stages* (Vol. II). Harper & Row.

Kohlberg, L. (1996). *Die Psychologie der Moralentwicklung.* Suhrkamp.

Kollmuss, A., & Agyeman, J. (2002). Mind the gap: Why do people act environmentally and what are the barriers to pro-environmental behavior? *Environmental Education Research, 8*(3), 239–260. https://doi.org/10.1080/13504620220145401

Krause, L.-K., & Gagné, J. (2021). *Die andere deutsche Teilung: Zustand und Zukunftsfähigkeit unserer Gesellschaft.* More in Common e.V. https://www.dieandereteilung.de/media/o5konmo3/more-in-common_fault-lines_executive-summary.pdf

Landmann, H. (2020). Emotions in the context of environmental protection: Theoretical considerations concerning emotion types, eliciting processes, and affect generalization. *Umweltpsychologie, 47*, 61–73. https://doi.org/10.31234/osf.io/yb2a7

Landmann, H., & Rohmann, A. (2020). Being moved by protest: Collective efficacy beliefs and injustice appraisals enhance collective action intentions for forest protection via positive and negative emotions. *Journal of Environmental Psychology, 71*, 101491. https://doi.org/10.1016/j.jenvp.2020.101491

Lanfranco, B., & Rava, C. (2014). Household demand elasticities for meat products in Uruguay. *Spanish Journal of Agricultural Research, 12*(1), 15. https://doi.org/10.5424/sjar/2014121-4615

Lewin, K. (1951). In D. Cartwright (Ed.), *Field theory in social science: Selected theoretical papers*. Harper & Row.

Lindenberg, S., & Steg, L. (2007). Normative, gain and hedonic goal frames guiding environmental behavior. *Journal of Social Issues, 63*(1), 117–137. https://doi.org/10.1111/j.1540-4560.2007.00499.x

Lubchenco, J., Cerny-Chipman, E. B., Reimer, J. J., & Levin, S. A. (2016). The right incentives enable ocean sustainability successes and provide hope for the future. *Proceedings of the National Academy of Sciences of the United States of America, 113*(51), 14507–14514. https://doi.org/10.1073/pnas.1604982113

Markowitz, E. M., & Shariff, A. F. (2012). Climate change and moral judgement. *Nature Climate Change, 2*(4), 243–247. https://doi.org/10.1038/nclimate1378

Martínez-Alier, J. (2014). The environmentalism of the poor. *Geoforum, 54*, 239–241. https://doi.org/10.1016/j.geoforum.2013.04.019

McCrae, R. R., & Costa, P. T. Jr. (1999). *A five factor theory of personality. Handbook of personality: Theory and research* (2. Auflage, pp. 139–153). Guilford Press.

McLeod, S. (2023, October 5). *Social identity theory simply psychology (Tajfel & Turner, 1979)*. Simply Psychology. https://www.simplypsychology.org/social-identity-theory.html

Meadows, D. (2008). *Thinking in systems: A primer*. Earthscan.

Michalke, A., Stein, L. C., Fichtner, R., Gaugler, T., & Stoll-Kleemann, S. (2022). True cost accounting in agri-food networks: A German case study on informational campaigning and responsible implementation. *Sustainability Science, 17*(6), 2269–2285. https://doi.org/10.1007/s11625-022-01105-2

Milfont, T. L., & Sibley, C. G. (2012). The big five personality traits and environmental engagement: Associations at the individual and societal level. *Journal of Environmental Psychology, 32*(2), 187–195. https://doi.org/10.1016/j.jenvp.2011.12.006

Montada, L., Schmitt, M., & Dalbert, C. (1986). *Thinking about justice and dealing with one's own privileges. Justice in social relations* (pp. 125–143). Springer.

Moser, S. C., & Dilling, L. (2011). Communicating climate change opportunities and challenges for closing the science action gap. *The Oxford handbook of climate change and society*. Oxford University Press.

Mozzato, D., Gatto, P., Defrancesco, E., Bortolini, L., Pirotti, F., Pisani, E., & Sartori, L. (2018). The role of factors affecting the adoption of environmentally friendly farming practices: Can geographical context and time explain the differences emerging from literature? *Sustainability, 10*(9), 3101. https://doi.org/10.3390/su10093101

Naito, R., Zhao, J., & Chan, K. M. A. (2022). An integrative framework for transformative social change: A case in global wildlife trade. *Sustainability Science, 17*(1), 171–189. https://doi.org/10.1007/s11625-021-01081-z

Norgaard, K. M. (2011). *Living in denial: Climate change, emotions, and everyday life*. MIT Press.

O'Connor, M. I., Mori, A., Gonzalez, A., Dee, L. E., Loreau, M., Avolio, M. L., Byrnes, J. E. K., Cheung, W., Cowles, J., Clark, A. T., Hautier, Y., Hector, A., Komatsu, K. J., Newbold, T., Outhwaite, C. L., Reich, P. B., Seabloom, E. W., Williams, L., Wright, A. J., & Isbell, F. (2021). Grand challenges in biodiversity–ecosystem functioning research in the era of science–policy platforms require explicit consideration of feedbacks. *Proceedings of the Royal Society, 288*(1960). https://doi.org/10.1098/rspb.2021.0783

O'Riordan, T., & Stoll-Kleemann, S. (2015). The challenges of changing dietary behavior toward more sustainable consumption. *Environment: Science and Policy for Sustainable Development, 57*(5), 4–13. https://doi.org/10.1080/00139157.2015.1069093

Painter, J. (2013). *Climate change in the media: Reporting risk and uncertainty.* Bloomsbury Publishing.

Parfit, D. (1987). *Reasons and persons* (Edition with corrections). Oxford University Press.

Peattie, K. J. (2010). Green consumption: Behavior and norms. *Annual Review of Environment and Resources, 35*(1), 195–228. https://doi.org/10.1146/annurev-environ-032609-094328

Peeters, W., Diependaele, L., Sterckx, S., McNeal, R. H., & De Smet, A. (2015). *Climate change and individual responsibility: Agency, moral disengagement and the motivational gap.* Palgrave MacMillian UK.

Piazza, J., Ruby, M. B., Loughnan, S., Luong, M., Kulik, J., Watkins, H. M., & Seigerman, M. (2015). Rationalizing meat consumption. The 4NS. *Appetite, 91,* 114–128. https://doi.org/10.1016/j.appet.2015.04.011

Piketty, T. (2022). *Eine kurze Geschichte der Gleichheit.* C.H. Beck.

Pinquart, M., & Silbereisen, R. (2007). Entwicklung des Umweltbewusstseins über die Lebensspanne. *Umweltpsychologie, 11*(7), 84–99.

Prokopy, L. S., Floress, K., Arbuckle, J. G., Church, S. P., Eanes, F. R., Gao, Y., Gramig, B. M., Ranjan, P., & Singh, A. S. (2019). Adoption of agricultural conservation practices in the United States: Evidence from 35 years of quantitative literature. *Journal of Soil and Water Conservation, 74*(5), 520–534. https://doi.org/10.2489/jswc.74.5.520

Rahmstorf, S. (2012). Is journalism failing on climate? *Environmental Research Letters, 7*(4), 041003. https://doi.org/10.1088/1748-9326/7/4/041003

Rau, H., Nicolai, S., Franikowski, P., & Stoll-Kleemann, S. (2024). Distinguishing between low- and high-cost pro-environmental behavior: Empirical evidence from two complementary studies. *Sustainability, 16,* 2206. https://doi.org/10.3390/su16052206

Rauschmayer, F., & Omann, I. (2012). Transition to sustainability: Not only big, but deep. *Gaia (Heidelberg, 21*(4), 266–268. https://doi.org/10.14512/gaia.21.4.7

Reese, G., & Jacob, L. E. (2015). Principles of environmental justice and pro-environmental action: A two-step process model of moral anger and responsibility to act. *Environmental Science & Policy, 51,* 88–94. https://doi.org/10.1016/j.envsci.2015.03.011

Ritson, C., & Petrovici, D. (2001). The economics of food choice: Is price important? *Food, people and society: A European perspective of consumers' food choices* (pp. 339–364). Springer.

Rodríguez-Entrena, M., & Arriaza, M. (2013). Adoption of conservation agriculture in olive groves: Evidences from Southern Spain. *Land Use Policy, 34,* 294–300. https://doi.org/10.1016/j.landusepol.2013.04.002

Roeser, S. (2012). Risk communication, public engagement, and climate change: A role for emotions. *Risk Analysis, 32*(6), 1033–1040. https://doi.org/10.1111/j.1539-6924.2012.01812.x

Roth, G. (2016). *Persönlichkeit, Entscheidung und Verhalten. Warum es so schwierig ist, sich und andere zu ändern: Persönlichkeit, Entscheidung und Verhalten.* Klett-Cotta.

Ruby, M. B., & Heine, S. (2011). Meat, morals, and masculinity. *Appetite, 56*(2), 447–450. https://doi.org/10.1016/j.appet.2011.01.018

Ryan, R. M., & Deci, E. L. (2000). Self-determination theory and the facilitation of intrinsic motivation, social development, and well-being. *American Psychologist, 55*(1), 68–78. https://doi.org/10.1037/0003-066x.55.1.68

Sabbagh, C., & Schmitt, M. (2016). Past, present, and future of social justice theory and research. *Handbook of social justice theory and research* (pp. 1–11). Springer.

Sachs, J., Kroll, C., Lafortune, G., Fuller, G., & Woelm, F. (2021). *Sustainable development report 2021*. Dublin University Press and Sustainable Development Solutions Network (SDSN). https://sdgtransformationcenter.org/reports/sustainable-development-report-2023

Saito, K. (2024). *Slow down: How degrowth communism can save the earth*. Orion Publishing Group.

Santoro, F., Scowcraft, G., Santin, S., Fauville, G., & Tuddenheim, P. (2017). Ocean literacy for all – A toolkit. *IOC manuals and guides 80*. IOC/UNESCO and UNESCO Venice Office.

Sapolsky, R. M. (2017). *Behave: The biology of humans at our best and worst*. Vintage/Penguin Random House.

Sawe, N. (2017). Using neuroeconomics to understand environmental valuation. *Ecological Economics, 135*, 1–9. https://doi.org/10.1016/j.ecolecon.2016.12.018

Sawe, N., & Chawla, K. (2021). Environmental neuroeconomics: How neuroscience can inform our understanding of human responses to climate change. *Current Opinion in Behavioral Sciences, 42*, 147–154. https://doi.org/10.1016/j.cobeha.2021.08.002

Schösler, H., De Boer, J., Boersema, J. J., & Aiking, H. (2015). Meat and masculinity among young Chinese, Turkish and Dutch adults in the Netherlands. *Appetite, 89*, 152–159. https://doi.org/10.1016/j.appet.2015.02.013

Schultz, P. W., & Stone, W. F. (1994). Authoritarianism and attitudes toward the environment. *Environment and Behavior, 26*(1), 25–37. https://doi.org/10.1177/0013916594261002

Schwartz, S. (2006). *Basic human values: An overview*. The Hebrew University of Jerusalem.

Schwartz, S. H. (1977). Normative influences on altruism. *Advances in experimental social psychology* (pp. 221–279). https://doi.org/10.1016/s0065-2601(08)60358-5

Schwerdtner Máñez, K., Born, W., & Stoll-Kleemann, S. (2024). Turning the tide: An analysis of factors influencing the adoption of biodiversity-enhancing measures on agricultural land at the German Baltic coast. *Sustainability, 16*(1), 317. https://doi.org/10.3390/su16010317

Serebrennikov, D., Thorne, F., Kallas, Z., & McCarthy, S. (2020). Factors influencing adoption of sustainable farming practices in Europe: A systemic review of empirical literature. *Sustainability, 12*(22), 9719. https://doi.org/10.3390/su12229719

Sexton, R. J. (2013). Market power, misconceptions, and modern agricultural markets. *American Journal of Agricultural Economics, 95*(2), 209–219. https://doi.org/10.1093/ajae/aas102

Sok, J., Hogeveen, H., Elbers, A. R. W., & Oude Lansink, A. G. J. M. (2016). Perceived risk and personality traits explaining heterogeneity in Dutch dairy farmers' beliefs about vaccination against bluetongue. *Journal of Risk Research, 21*(5), 562–578. https://doi.org/10.1080/13669877.2016.1223162

Southerton, D. (2013). Habits, routines and temporalities of consumption: From individual behaviours to the reproduction of everyday practices. *Time & Society, 22*(3), 335–355. https://doi.org/10.1177/0961463x12464228

Stanley, S. K., Hogg, T. L., Leviston, Z., & Walker, I. (2021). From anger to action: Differential impacts of eco-anxiety, eco-depression, and eco-anger on climate action and wellbeing. *The Journal of Climate Change and Health, 1,* 100003. https://doi.org/10.1016/j.joclim.2021.100003

Steg, L., Bolderdijk, J. W., Keizer, K. E., & Perlaviciute, G. (2014). An integrated framework for encouraging pro-environmental behavior: The role of values, situational factors and goals. *Journal of Environmental Psychology, 38,* 104–115. https://doi.org/10.1016/j.jenvp.2014.0.002

Stengel, O. (2011). *Suffizienz: Die Konsumgesellschaft in der ökologischen Krise* [PhD diss., Universität Jena]. https://epub.wupperinst.org/frontdoor/index/index/docId/3822

Stern, P. C. (2000). New environmental theories: Toward a coherent theory of environmentally significant behavior. *Journal of Social Issues, 56*(3), 407–424. https://doi.org/10.1111/0022-4537.00175

Stiglitz, J. E. (2013). *The price of inequality: How Today's divided society endangers our future.* W. W. Norton & Company.

Stoll-Kleemann, S. (2019). Feasible options for behavior change toward more effective ocean literacy: A systematic review. *Frontiers in Marine Science, 6*(237). https://doi.org/10.3389/fmars.2019.00273

Stoll-Kleemann, S., & Schmidt, U. (2017). Reducing meat consumption in developed and transition countries to counter climate change and biodiversity loss: A review of influence factors. *Regional Environmental Change, 17*(5), 1261–1277. https://doi.org/10.1007/s10113-016-1057-5

Stoll-Kleemann, S., O'Riordan, T., & Jaeger, C. (2001). The psychology of denial concerning climate mitigation measures: Evidence from Swiss focus groups. *Global Environmental Change, 11*(2), 107–117. https://doi.org/10.1016/s0959-3780(00)00061-3

Stoll-Kleemann, S., Nicolai, S., & Franikowski, P. (2022). Exploring the moral challenges of confronting high-carbon-emitting behavior: The role of emotions and media coverage. *Sustainability, 14*(10), 5742. https://doi.org/10.3390/su14105742

Swearingen, T. C. (1990). *Moral development and environmental ethics.* [Dissert. Abs. Int. 50 (12-B, Part 1), 5905, University of Washington]. https://philpapers.org/rec/SWEMDA-2

Thaler, R. H., & Sunstein, C. R. (2022). *Nudge: Improving decisions about health, wealth and happiness.* Penguin UK.

Tillack, H. (2015). *Die Lobby-Republik: Wer in Deutschland die Strippen zieht.* Hanser.

United Nations Framework Convention on Climate Change (UNFCCC). (2016). *Adoption of the Paris Agreement.* United Nations.

Van Zomeren, M., Postmes, T., & Spears, R. (2008). Toward an integrative social identity model of collective action: A quantitative research synthesis of three sociopsychological perspectives. *Psychological Bulletin, 134*(4), 504–535. https://doi.org/10.1037/0033-2909.134.4.504

Vartanian, L. R. (2015). Impression management and food intake. Current directions in research. *Appetite, 86,* 74–80. https://doi.org/10.1016/j.appet.2014.08.021

Verain, M., Dagevos, H., & Antonides, G. (2015). Flexitarianism: A range of sustainable food styles. *Handbook of research on sustainable consumption* (pp. 209–223). Edward Elgar Publishing.

Verplanken, B., & Roy, D. (2016). Empowering interventions to promote sustainable lifestyles: Testing the habit discontinuity hypothesis in a field experiment. *Journal of Environmental Psychology, 45*, 127–134. https://doi.org/10.1016/j.jenvp.2015.11.008

Whitley, S., van der Burg, L., Worrall, L., & Patel, S. (2017). *Cutting Europe's lifelines to coal: Tracking subsidies in 10 countries*. Overseas Development Institute.

Wissenschaftlicher Beirat der Bundesregierung Globale Umweltveränderungen (WBGU). (2018). *Welt im Wandel: Gesellschaftsvertrag für eine große Transformation* (2nd ed.). https://www.wbgu.de/fileadmin/user_upload/wbgu/publikationen/hauptgutachten/hg2011/pdf/wbgu_jg2011.pdf (originally published 2011).

Withana, S., ten Brink, P., Franckx, L., Hirschnitz-Garbers, M., Mayeres, I., Oosterhuis, F., & Porsch, L. (2012). *Study supporting the phasing out of environmentally harmful subsidies: A report by the institute for European environmental policy (IEEP), final report*. Institute for Environmental Studies – Vrije Universiteit (IVM), Ecologic Institute and VITO for the European Commission – DG Environment.

4 Justifying Climate-Unjust Individual Behavior

Barriers to Climate Action as Moral Disengagement and Other Forms of Justification

Susanne Nicolai

Contradictory perspectives emerge in the face of climate injustice. On the one hand, people genuinely have a fundamental motive to restore justice, which can be considered as part of the psychological needs for consistency and relatedness (see Chapter 3). On the other hand, climate injustice is the reality and becomes increasingly obvious with the ongoing climate crisis.

Based on the referent cognition theory by Folger (1986, see Chapter 1), injustice is associated with negative emotions due to the gap between perceived reality and imaginary alternatives. The appraisal of the negative emotion is therefore based on this discrepancy. Festinger (1957) provides support for this postulated process with his pioneering theory on cognitive dissonance. The scientific literature confirms that people experience dissonance when they become aware that their behavior may not match their values, attitudes or knowledge. This dissonance is difficult to bear and feels highly uncomfortable. The resulting psychological discomfort is able to motivate people to use inconsistency compensation strategies to solve the stressful state.

Two kinds of inconsistency compensation strategies exist: first, direct inconsistency compensation strategies, such as behavior change (i.e. become a vegetarian), attitude change (i.e. thinking that individual behavior will not make a difference) or behavior modification (i.e. reduce meat consumption to a minimum); and second, indirect inconsistency compensation strategies, comprising justification mechanisms like rationalization (i.e. arguing that one needs meat out of health reasons) or non-compensation (i.e. passively forgetting the inconsistency; Festinger, 1957). These strategies vary in their need for elaboration and effort. As Festinger's (1957) groundbreaking studies were able to show, behavior change is the most effortful compensation strategy, followed by behavior modification, rationalization, trivialization, attitude change and non-compensation (Köhler et al., 2019; Leippe & Eisenstadt, 1999). Justification mechanisms and attitude change are therefore an easy, short-term solution to avoid the unpleasant feelings resulting from cognitive dissonance.

DOI: 10.4324/9781003033547-4

Bandura (2002, 2016) further built on Festinger (1957) and developed the theory of moral disengagement (in simple words "How people do harm and still live with themselves", as the book title says). Moral disengagement is the psychological process through which individuals or groups detach themselves from conventional (or their own) ethical standards of behavior, subsequently convincing themselves that engaging in new, unethical actions can be justified, often attributing their actions to perceived extenuating circumstances. The more comprehensive perspective of moral disengagement views moral action as a result of the complex interplay between cognitive, emotional and social influences, with personal agency operating within a broader network of sociostructural factors. This approach involves separating moral reactions from inhumane conduct, leading people to believe that ethical standards may not apply in certain contexts (Bandura, 2002). These justifications can therefore be seen as a way of cognitive restructuring as people find situational exceptions of their moral values.

Bandura's moral disengagement strategies arise at both the individual and social-systems level. He categorized eight psychosocial mechanisms (see Figure 4.1). The initial three mechanisms operate when individuals transform harmful practices into acceptable ones through (1) moral justification, which involves framing inappropriate behavior as necessary for other moral values (i.e. to protect friends or family), (2) advantageous comparison, where individuals compare their actions favorably to others with a higher carbon dioxide (CO_2) footprint and (3) euphemistic labeling, in which harmful actions are given less alarming descriptions to make them appear harmless. Through two additional mechanisms – (4) displacement and (5) diffusion of responsibility – individuals absolve themselves of personal accountability by shifting the blame onto others. Displacing the responsibility means directly transferring it to another person or institution, whereas diffusion means that no one really knows who is responsible. The diffusion of responsibility further allows suspending moral control by subdividing activities that may seem harmless in isolation.

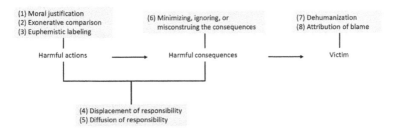

Figure 4.1 The eight mechanisms of moral disengagement by Bandura (1986). Own representation

Another strategy – (6) minimizing, ignoring or misconstruing consequences – aims to reinterpret the harmful effects of actions to reduce guilt or remorse. Climate change denial falls under this category. The final two mechanisms – (7) marginalizing and (8) blaming the victims – involve attributing responsibility for worsening ecological conditions to the victims themselves.

According to Moore (2015), it makes sense to follow Bandura (1986) in treating these eight moral disengagement mechanisms "as a coherent set of cognitive tendencies that influence the way individuals may approach decisions with ethical import" (Moore, 2015, p. 6). However, others have discussed or studied similar cognitive mechanisms separately (e.g. euphemistic language, diffusion of responsibility, exonerative comparison).

Stoll-Kleemann et al. (2001) also found these justification strategies in the environmental sphere in a Swiss sample. Here, again these justification mechanisms were used to "assuage guilt, to reinforce victim status, to justify resentment or anger, and to emphasize the negative feelings toward disliked behavior (e.g. the disagreeable qualities of relying on public transport and the loss of social prestige involved)" (Stoll-Kleemann et al., 2001, p. 112). In the following paper, justification strategies in the environmental domain were categorized according to Bandura's theory of moral disengagement. The first and most widely used mechanism was the displacement of responsibility, followed by moral justifications, and later disregard, distortion or denial of harmful effects (Stoll-Kleemann & O'Riordan, 2020). The most effective strategies are mechanisms that operate at the behavior locus: People transform harmful practices into worthy ones through social and moral justification, exonerative social comparison and sanitizing language (Bandura, 2007).

In a recent paper, we (Stoll-Kleemann et al., 2023) related the concept of moral disengagement to climate-just behavior and developed a questionnaire to make these justifications measurable in quantitative studies. In the process of developing the questionnaire (Moral Disengagement in High Carbon Behavior, MD-HCB), we changed the construct dehumanization to that of the less harsh social distance. Inspired by Markowitz and Shariff (2012), we also added the category of the blamelessness of unintentional action to Bandura's (1986) eight mechanisms of moral disengagement. With this ninth component, both versions of our questionnaire display good reliability and despite its abbreviated length, the short version also manifests sound diagnostic properties. The questionnaire is measured with the following items (see Table 4.1).

In the following studies, moral disengagement in high-carbon behavior showed to be the strongest predictor (in a set of various predictors) of both pro-environmental intention and behavior (Nicolai et al., 2022). This stands in line with other current literature. Leviston and Walker (2020) indicate that moral engagement levels play a mediating role in the relationship between (1) beliefs about the causes of climate change and pro-environmental behavior, (2) individual perceptions of response efficacy and pro-environmental

Table 4.1 Questionnaire to measure the tendency to morally disengage in high-carbon behavior (MD-HCB) from Stoll-Kleemann et al. (2023)

Moral disengagement in high carbon behavior questionnaire (MD-HCB)

Moral justification

It is all right to have a high carbon footprint if it is advantageous for me or for my friends and relatives.

Not talking about my high CO_2 emissions is justified if it gives a better impression of my friends and myself.

Euphemistic labeling

Not mentioning the negative effects of climate change is OK, as long as the personal benefits you derive from a lifestyle that harms the climate outweigh them.

Until technical solutions for activities that damage the climate have been found, it is all right to pursue them.

Advantageous comparison

I think the CO_2 emissions that I am personally responsible for – even if they are higher than the global average – are less of a reason for me to be concerned than those produced by rich people, business and industry, and other countries.

If you look at the CO_2 emissions levels of countries like the USA and China, here in Germany we do not have to worry so much about our own.

Displacement of responsibility

Producing a high level of CO_2 emissions is acceptable if your friends push you to do the things that cause them, like eating meat, driving a lot or taking plane trips.

Individuals should not be held personally responsible for their own high CO_2 emissions levels because at the end of the day, it is the politicians whose decisions have created the situation.

Diffusion of responsibility

I should not be held personally responsible for my high CO_2 emissions because most other people produce levels that are just as high, and given the overall amount, my behavior makes very little difference.

Lying about a high CO_2 emissions level is ok if my friends think it is better to do it.

Distortion of consequences

CO_2 emissions that are slightly above average do not cause a great deal of damage.

behavior and (3) perceived responsibility for contributing to climate change and pro-environmental behavior. Additionally, regardless of their beliefs about the causes of climate change, individuals tend to attribute more responsibility to groups and organizations and less to individuals for both causing and addressing climate change. Analyzing data over time revealed that individuals who experienced a decrease in moral engagement also reported a decrease in feelings of guilt related to climate change, with indications that this connection operates bidirectionally.

Furthermore, moral disengagement in high-carbon behavior was associated with victim sensitivity (Nicolai et al., 2022). We therefore can conclude that people who easily perceive themselves as being the victim of a situation also tend to justify their climate-unjust behavior more. This is in line with current literature on victim sensitives who often throw their moral standards overboard as they fear to be disadvantaged otherwise. Individual differences (as justice sensitivity) consequently also play a role in the amount of justification of climate-unjust behavior.

However, moral disengagement is able to explain why individuals do not act in line with their intentions (intention-behavior gap; Sheeran & Webb, 2016) or their knowledge (knowledge-behavior-gap; Sligo & Jameson, 2000). Moreover, although pro-environmental attitude is a predictor of pro-environmental behavior, individuals do not always behave in line with their attitudes about a situation or problem, called attitude-behavior gap (see Chapter 5; Bamdad, 2019). For example, Farjam et al. (2019) showed that attitude predicts pro-environmental behavior in low-cost behavior but not in high-cost behavior.

It is important to note here is that moral disengagement not only occurs at the individual level but also the collective one. Bandura himself mentioned that moral disengagement at the collective level can have detrimental effects by "supporting, justifying, and legitimizing inhumane social practices and policies" (Bandura et al., 1996, p. 372). Feygina et al. (2010) conclude that climate change denial (one of the eight mechanisms of Bandura, 1986) is a form of system justification, especially endorsed by those who support capitalism and the status quo. This stands in line with the system justification theory, which proposes that the need for safety, certainty and stability "gives rise to a motivation to perceive the system as fair, legitimate, beneficial, and stable, as well as the desire to maintain and protect the status quo" (Feygina et al., 2010, p. 327). While system justification can trigger a rationalization process to accept the existing state of affairs and aid individuals in coping with undesirable realities, it may also hinder the formation of intentions or the initiation of actions aimed at addressing injustices or systemic issues. People who take advantage of the current system rather tend to justify the system, but also people who are disadvantaged justify the system due to these mechanisms. For instance, individuals with limited financial means frequently do not view the immense wealth of the super-rich as inherently unjust. Instead, they aspire to attain a similar status, influenced by the conviction that diligent effort can elevate anyone to super-rich status (Kronenberg, 2023; Walasek & Brown, 2016).

The dynamics of individual interaction within a system are elucidated by multi-level models (e.g. Göpel, 2016) that examine how socio-technical societal subsystems interact during transformation processes across temporal and spatial dimensions. These models delineate three main levels: the macro-level, encompassing overarching phenomena such as megatrends like climate change, the market system and hegemonic paradigms; the meso-level, comprising regimes such as policy, technology and science; and the micro-level, represented by niches (Wullenkord & Hamann, 2021). The higher levels – characterized by institutionalization, inertia and historical roots – wield significant influence but are resistant to rapid change. Regimes, entrenched by path-dependencies like institutionalization and social-psychological infrastructures (e.g. norms, shared beliefs; see Welzer, 2011), impede the envisioning of alternatives, maintain the status quo and hinder swift transformations. Conversely, niches provide protected spaces where radical sociotechnical innovations can be tested. When external pressures such as climate change destabilize regimes, windows of opportunity emerge, facilitating the establishment of niche innovations within these regimes. Individuals – both individually and within their groups, as well as at the regime level –disengage from responsibility in response to these dynamics (Wullenkord & Hamann, 2021).

Göpel (2016) explicitly acknowledges individuals and hegemonic paradigms in transformations by adding two layers: The mini-level contains individuals making up institutions. The meta-level represents the "hegemonic paradigm and common sense framework that serves as a reference for individual strategies and narratives" of change (Göpel, 2016, p. 47). Both levels interact: The mini-level influences the meta-level because every individual contributes to changing and shaping the future paradigm, and thereby reality. The meta-level is deeply embedded in the meso-, micro-, and mini-levels and mediates between them. For instance, it affects how individuals in specific regimes think (cognitive lock-ins; see Welzer, 2011; Wullenkord & Hamann, 2021).

Considerable scientific support comes from the research on the belief in a just world. The belief in a just world describes the conviction that the world is fundamentally fair and that everyone ultimately gets what they deserve and again results from the *justice motive* (see Chapter 2) due to the need for safety and stability (Dalbert, 1992; Lerner, 1980). Both the justice motive and the resulting belief in a just world are conceived as interindividually varying personality traits depending on socialization. In combination with experiences of helplessness and low self-efficacy, a strong belief in a just world can contribute to devaluing victims, attributing to them a share of the blame for their fate and trivializing or justifying disadvantages (victimization, secondary). Through this cognitive reinterpretation, cognitive dissonance is avoided and the belief in a just world as well as the associated emotional security is maintained. In combination with perceived opportunities for action, a strongly developed just world belief can lead to compensatory support for victims of observed injustice to restore justice.

As Feygina et al. (2010) argue, the environmental peril we confront arises from the very structure of the status quo; our socioeconomic practices have led to the present crisis, making it an internal (or endogenous) threat to the system. Confronting this type of threat entails: (1) recognizing the deficiencies in the existing system and established norms, (2) embracing collective responsibility at both systemic and individual levels for the current state of the environment, and (3) acknowledging that the status quo must transform to avert ecological catastrophe. Feygina et al. (2010) have gathered compelling evidence establishing a direct link between system justification tendencies and the denial of environmental issues, as well as the reluctance to participate in pro-environmental actions. These pioneering findings offer the initial empirical confirmation that system justification is connected to denial in terms of problems inherent within the social system. Additionally, they underscore the significant role collective denial (as one of the moral disengagement strategies) plays in perpetuating an unfavorable status quo.

Through their own experiment, Feygina et al. (2010) demonstrated how this insight can be applied to political communication. Participants exposed to a preservation framing, which conveyed climate change as the foremost threat to our safety, exhibited reduced tendencies toward system justification compared to a control group without framing. Consequently, they displayed greater intentions to engage in pro-environmental behavior and were more inclined to support pro-environmental initiatives, such as signing petitions (Feygina et al., 2010). This approach offers a potential remedy for addressing the inhibiting effects of moral disengagement at a collective level, highlighting that alternatives do not necessarily entail safety risks. On the contrary, climate-just alternatives safeguard the essential elements of human existence. Given humans' inherent inclination toward risk aversion (Beaud & Willinger, 2015), this approach could represent the future direction of climate change communication, addressing the psychological need for safety.

We propose that moral disengagement operates not only across different societal and individual levels but also across various levels of cognitive processing. According to the theory of cognitive dissonance, the management of unpleasant emotions typically follows the path of least resistance, relying on justification mechanisms rather than behavioral change, depending on available cognitive resources. This might bring up various forms of moral disengagement: For individuals with limited engagement in the climate crisis, denial of its existence, and avoidance of related news may represent the easiest coping strategy to mitigate upcoming stress. Conversely, individuals who are more familiar with the climate crisis may find it easier to engage in downward social comparisons (e.g. comparison to people who take a plane more than once a year).

Batzke and Cohrs (2020) did not specifically examine moral disengagement but rather investigated justifications for not engaging in climate-just behavior in general. They suggest that the crucial question is not whether one justifies and to what extent, but how. According to the authors, there are two types of justifications: recognition justifications, which involve admitting guilt and promising improvement (although behavior remains unchanged afterward, making these forms of justification somewhat hypocritical), and defense justifications such as asserting necessity, individual powerlessness or appealing to convenience. However, in an experimental setting, Batzke and Cohrs (2020) were unable to demonstrate any differences in their wish to behave more pro-environmentally between groups who were nudged to use one of these justification categories each.

Consequently, successful interventions to minimize moral disengagement are missing yet necessary, given that people tend to avoid or resist information about the negative consequences of their non-climate-just behavior because they contradict or threaten their basic perspectives on fairness and ethical behavior (Stoll-Kleemann et al., 2001). Because people who use moral disengagement are unlikely to change their behavior toward climate-just behavior (Nicolai et al., 2022), it is important to look at these mechanisms to find solutions to overcoming them.

If such interventions are missing, moral disengagement of the majority may lead to climate silence by those, who are already aware of climate change consequences (Heald, 2017). Climate silence refers to the phenomenon of individuals or organizations refraining from discussing or addressing climate change, often due to fear of negative consequences or a lack of awareness of others' concerns (Civelek et al., 2015; Geiger & Swim, 2016). This silence can be perpetuated by climate change contrarianism in the media, which can hinder efforts to address the issue (Brisman, 2012). Additionally, the framing of climate-related crises as solely caused by climate events can silence the social and political-economic causes, leading to a narrow understanding of the problem (Ribot, 2022).

A significant role in shaping societal discourses, and, therefore, counteracting climate silence, has the media. The media is a crucial source of information for people's awareness and knowledge about climate change (Schäfer, 2015). The role of the media in climate change communication is outstanding, as it can influence public awareness, attitudes, and knowledge of climate change (Ha et al., 2019). However, its effectiveness varies. For example, the principle of covering different perspectives on topics in journalism gave climate change deniers a very successful platform for years, although scientific consensus about human-made climate change had long been certain (see Chapter 3).

Moreover, lobbyism in both media and politics hinders progress. Regarding the media, the fossil lobby has been found to have engaged in disinformation campaigns to downplay the severity of climate change and its link to human activity (Franta, 2021; Mulvey et al., 2015; Supran & Oreskes, 2017). These campaigns have been used to promote policies favorable to the fossil fuel industry and cast doubt on the scientific consensus on climate change. The tactics employed include the use of corporate front groups, public relations firms and conservative think tanks to spread misinformation and delay climate action (Beder, 2011; see Chapter 3).

The current Global Risk Report by the World Economic Forum (2024) indicates misinformation, especially with new opportunities of AI-generated pictures, videos and texts, as the highest global risk (followed by extreme weather events). Chan et al. (2017) highlight that "because misinformation can lead to poor decisions about consequential matters and is persistent and difficult to correct, debunking it is an important scientific and public-policy goal" (p. 1531). In their meta-analyses, they ascertain three recommendations for successful debunking of false information. First, the generation of arguments in line with the misinformation should be reduced. Second, conditions that facilitate scrutiny and counter arguing of misinformation should be created. And third, misinformation should be corrected with new detailed information. Yet another discovery from the meta-analyses is regrettably that this method does not consistently succeed. Particularly on social media platforms, where sensationalist content like misinformation is frequently disseminated due to its provocative nature, alternative posts presenting accurate information may not garner the same level of attention, as they lack novelty and provocativeness.

Regarding climate politics, lobbying exerts a substantial influence. A study conducted by Meng and Rode (2019) quantifies this impact, showing that political lobbying reduced the likelihood of enacting the Waxman-Markey bill (draft law to reduce carbon emissions) in the United States by 13 percentage points, equating to US$60 billion in projected climate damages. This shows the huge amount of social costs that follow lobbyism of large industries. The Waxman-Markey bill was not passed. As Bernhagen (2019) argues, lobbying has significantly increased in recent decades and is becoming increasingly professional. Lobbying strategies range from discussions at elite levels to media-savvy mass protests. The success of lobbying efforts depends particularly on the politically relevant information and the quality of the lobbyists' contact networks. These resources are unevenly distributed in society, contradicting the principles of democratic governance. Effective regulation could provide a remedy but is lacking in most political systems (see Chapter 3).

Lobbyism benefits from collective moral disengagement, as individuals fail to monitor and question political measures, refrain from engaging in meaningful discourse in the media and generally abstain from involvement in societal processes. Since resources for lobbying are unevenly distributed, this is a danger to democracies. To regulate lobbyism, Johnson (2006) proposes that two types of regulation laws exist: prohibitions and disclosure requirements (also called sunshine laws). As prohibitions are more effective, Johnson (2006) takes a closer look on various prohibition strategies. For example, the prohibition of false statements is difficult, but possible, similar to laws in consumer protection. Furthermore, gifts, meals, entertainment, and travel offers by lobbyists can be prohibited. Moreover, lobbyist branches could be given a contingent. Climate activists could then receive the same amount of time in a politician's office as the fossil lobby, for example. After this deeper look inside many different prohibition strategies, Johnson (2006) concludes:

> The interesting thing is that if one looks at the law nationally, for virtually every issue that someone can identify, some legislature, somewhere, has passed a rule that effectively addresses the problem, or has taken action that would assist reformers in crafting appropriate solutions. Thus, the challenge in regulating lobbyists is not to re-conceptualize the field or to develop radically innovative solutions, but to employ the tools that are already available.

(p. 160)

However, not only top-down regulations such as prohibitions exist, but also bottom-up initiatives: Deighton-Smith (2004) substantiates transparency laws and initiatives in overcoming lobbyism and corruption. NGOs emphasize transparency as a fundamental element of civil society, enabling citizen empowerment. Against the backdrop of diminishing trust in governments, transparency emerges as a potent antidote to this erosion of trust. It compels governments to uphold elevated standards of behavior by ensuring their actions are subject to public scrutiny. By providing stakeholders with visibility into government actions and decisions, transparency fosters trust by enabling informed judgment of their quality. However, it can be frustrating and time intense to ask authorities for specific documents. Therefore, in Germany, an NGO (*FragDenStaat*) wrote a beginner guide and helped to submit the request to an authority (FragDenStaat, n.d.).

Furthermore, positive examples exist in media: a current study has shown that local media outlets can increase public engagement by working with local academics and scientists to communicate the local impact of climate change (Howarth & Anderson, 2019). The media is able to explain complex topics to everyone and bring them closer. Therefore, it also has the responsibility to do so regarding topics that affect societies around the globe.

These proposals of better climate communication and more transparency seem promising to reduce moral disengagement both individually and collectively as they give many concrete examples to become active and therefore make behavior change an easier way. Returning to the theory of cognitive dissonance from the beginning of the chapter, the management of unpleasant emotions typically follows the path of least resistance, depending on available cognitive resources. When people are confronted with *problem knowledge*, meaning information about climate change and its consequences, it therefore might also help them to obtain *action knowledge*, meaning information about what can be done concretely. Action knowledge may include specific ways of pro-environmental behaviors (individual and collective ones) as well as the impact of various pro-environmental behaviors. It then can be developed as a personal and collective competence, which was found to be crucial for taking action for the environment (Chawla & Cushing, 2007; Liobikienė & Poškus, 2019).

At the same time, action knowledge is needed to develop a feeling of self-efficacy (the belief in one's ability to perform well in a specific task). Research consistently shows that self-efficacy plays a crucial role in promoting pro-environmental behavior (see also Chapter 3). Hamann and Reese (2020) showed a positive relationship between self-, collective- and participatory efficacy and pro-environmental behavior. Collective efficacy is considered the belief of group members in the group's ability to reach desired goals (Bandura, 1997). Collective efficacy was successfully increased by giving participants action knowledge about the impact of various collective pro-environmental behaviors (Jugert et al., 2016; Van Zomeren et al., 2010).

Participatory efficacy describes the belief that an individual actor can make a significant difference in attaining a group's goal (Hamann & Reese, 2020; Van Zomeren et al., 2013).

Knowing what to do and the belief that one can make a difference (individually and in a group), therefore, are a key in reducing moral disengagement strategies.

References

Bamdad, T. (2019). Pro-environmental attitude-behavior: A spillover or a gap? *Advances in science, technology & innovation* (pp. 169–183). https://doi.org/10.1007/978-3-030-10804-5_17

Bandura, A. (1986). *social foundations of thought and action: A social cognitive theory.* Prentice Hall.

Bandura, A. (1997). *Self-efficacy: The exercise of control.* Freeman.

Bandura, A. (2002). Selective moral disengagement in the exercise of moral agency. *Journal of Moral Education, 31*(2), 101–119. https://doi.org/10.1080/0305724022014322

Bandura, A. (2007). Impeding ecological sustainability through selective moral disengagement. *International Journal of Innovation and Sustainable Development, 2*(1), 8–35. https://doi.org/10.1504/ijisd.2007.016056

Bandura, A. (2016). *Moral disengagement: How people do harm and live with themselves.* Worth Publishers.

Bandura, A., Barbaranelli, C., Caprara, G. V., & Pastorelli, C. (1996). Mechanisms of moral disengagement in the exercise of moral agency. *Journal Of Personality and Social Psychology, 71*(2), 364–374. https://doi.org/10.1037/0022-3514.71.2.364

Batzke, M., & Cohrs, C. (2020). "Ja, aber…" – Förderung klimafreundlichen Verhaltens durch die Veränderung individueller Rechtfertigungsstrategien. Kurzbericht ["Yes, but … "– Promoting environmentally Friendly behavior by altering individual account strategies. Short report]. *Umweltpsychologie, 24*(1), 131–141. https://www.research-gate.net/publication/348331494_Ja_aber_-_Forderung_klimafreundlichen_Verhaltens_durch_die_Veranderung_individueller_Rechtfertigungsstrategien_Kurzbericht

Beaud, M., & Willinger, M. (2015). Are people risk vulnerable? *Management Science, 61*(3), 624–636. https://doi.org/10.1287/mnsc.2013.1868

Beder, S. (2011). Lobbying, greenwash and deliberate confusion: How vested interests undermine climate change regulation. In *Conference for green thought and environmental politics*, Institute of European and American Studies. https://www.researchgate.net/publication/275276392_Lobbying_Greenwash_and_Deliberate_Confusion_How_Vested_Interests_Undermine_Climate_Change_Regulation

Bernhagen, P. (2019). *Lobbyismus in der Politikberatung* (pp. 249–261). Springer eBooks. https://doi.org/10.1007/978-3-658-03483-2_23

Brisman, A. (2012). The cultural silence of climate change contrarianism. *Climate change from a criminological perspective* (pp. 41–70). Springer New York.

Chan, M. S., Jones, C. R., Jamieson, K. H., & Albarracín, D. (2017). Debunking: A meta-analysis of the psychological efficacy of messages countering misinformation. *Psychological Science, 28*(11), 1531–1546. https://doi.org/10.1177/0956797617714579

Chawla, L., & Cushing, D. F. (2007). Education for strategic environmental behavior. *Environmental Education Research, 13*(4), 437–452. https://doi.org/10.1080/13504620701581539

Civelek, M. E., Aşçı, M. S., & Çemberci, M. (2015). Identifying silence climate in organizations in the framework of contemporary management approaches. *DOAJ*. https://doaj.org/article/f1be4e0f5e9d4c2bb0b3cbaae6270aa7

Dalbert, C. (1992). Der Glaube an die gerechte Welt: Differenzierung und Validierung eines Konstrukts. *Zeitschrift für Sozialpsychologie, 23*(4), 268–276. https://psycnet.apa.org/record/1993-86269-001

Deighton-Smith, R. (2004). Regulatory transparency in OECD countries: Overview, trends and challenges. *Australian Journal of Public Administration, 63*(1), 66–73. https://doi.org/10.1111/j.1467-8500.2004.00360.x

Farjam, M., Nikolaychuk, O., & Bravo, G. (2019). Experimental evidence of an environmental attitude-behavior gap in high-cost situations. *Ecological Economics, 166*, 106434. https://doi.org/10.1016/j.ecolecon.2019.106434

Festinger, L. (1957). A theory of cognitive dissonance. Stanford University Press eBooks. https://doi.org/10.1515/9781503620766

Feygina, I., Jost, J. T., & Goldsmith, R. E. (2010). System justification, the denial of global warming, and the possibility of "System-sanctioned change." *Personality and Social Psychology Bulletin, 36*(3), 326–338. https://doi.org/10.1177/01461672093 51435

Folger, R. (1986). *Rethinking equity theory: A referent cognitions model. Justice in social relations.* 145–162. Springer US.

FragDenStaat. (n.d.). *Hilfe und Tipps zum Einstieg auf FragDenStaat.de.* FragDenStaat. https://fragdenstaat.de/informationsfreiheit/einsteiger-guide/

Franta, B. (2021). Early oil industry disinformation on global warming. *Environmental Politics, 30*(4), 663–668. https://doi.org/10.1080/09644016.2020.1863703

Geiger, N., & Swim, J. K. (2016). Climate of silence: Pluralistic ignorance as a barrier to climate change discussion. *Journal of Environmental Psychology, 47*, 79–90. https://doi.org/10.1016/j.jenvp.2016.05.002

Göpel, M. (2016). *The great mindshift: How a new economic paradigm and sustainability transformations go hand in hand.* Springer.

Ha, J., Akhtar, R., Masud, M. M., Rana, S., & Banna, H. (2019). The role of mass media in communicating climate science: An empirical evidence. *Journal of Cleaner Production, 238*, 117934. https://doi.org/10.1016/j.jclepro.2019.117934

Hamann, K. R. S., & Reese, G. (2020). My influence on the world (of others): Goal efficacy beliefs and efficacy affect predict private, public, and activist pro-environmental behavior. *Journal Of Social Issues, 76*(1), 35–53. https://doi.org/10.1111/josi.12369

Heald, S. G. (2017). Climate silence, moral disengagement, and self-efficacy: How Albert Bandura's theories inform our climate-change predicament. *Environment: Science and Policy for Sustainable Development, 59*(6), 4–15. https://doi.org/10.1080/00139157.2017.1374792

Howarth, C., & Anderson, A. (2019). Increasing local salience of climate change: The un-tapped impact of the media-science interface. *Environmental Communication, 13*(6), 713–722. https://doi.org/10.1080/17524032.2019.1611615

Johnson, V. R. (2006). Regulating lobbyists: Law, ethics, and public policy. *Cornell Journal of Law and Public Policy, 16*(1), 1–56. https://commons.stmarytx.edu/cgi/viewcontent.cgi?article=1415&context=facarticles

Jugert, P., Greenaway, K. H., Barth, M., Büchner, R., Eisentraut, S., & Fritsche, I. (2016). Collective efficacy increases pro-environmental intentions through increasing self-efficacy. *Journal of Environmental Psychology, 48*, 12–23. https://doi.org/10.1016/j.jenvp.2016.08.003

Köhler, J., Geels, F. W., Kern, F., Markard, J., Wieczorek, A., Alkemade, F., Avelino, F., Bergek, A., Boons, F., Fünfschilling, L., Hess, D. J., Holtz, G., Hyysalo, S., Jenkins, K., Kivimaa, P., Martiskainen, M., McMeekin, A., Mühlemeier, M. S., Nykvist, B., & Wells, P. E. (2019). An agenda for sustainability transitions research: State of the art and future directions. *Environmental Innovation and Societal Transitions, 31,* 1–32. https://doi.org/10.1016/j.eist.2019.01.004

Kronenberg, M. (2023, February 11). Erde an Robin Hood: Bitte kommen [Essay]. *Steady – Treibhauspost.* https://steadyhq.com/de/treibhauspost/posts/27b74efd-e13c-4b9e-a7e1-ce75e7030cfc

Leippe, M. R., & Eisenstadt, D. (1999). A self-accountability model of dissonance reduction: Multiple modes on a continuum of elaboration. In E. Harmon-Jones & J. Mills (Eds.), *Cognitive dissonance: Progress on a pivotal theory in social psychology* (pp. 201–232). American Psychological Association. https://doi.org/10.1037/10318-009

Lerner, M. J. (1980). *Belief in a just world: A fundamental delusion.* Plenum Publishing Corporation.

Leviston, Z., & Walker, I. (2020). The influence of moral disengagement on responses to climate change. *Asian Journal of Social Psychology, 24*(2), 144–155. https://doi.org/10.1111/ajsp.12423

Liobikienė, G., & Poškus, M. S. (2019). The importance of environmental knowledge for private and public sphere pro-environmental behavior: Modifying the value-belief-norm theory. *Sustainability, 11*(12), 3324. https://doi.org/10.3390/su11123324

Markowitz, E. M., & Shariff, A. F. (2012). Climate change and moral judgement. *Nature Climate Change, 2*(4), 243–247. https://doi.org/10.1038/nclimate1378

Meng, K. C., & Rode, A. (2019). The social cost of lobbying over climate policy. *Nature Climate Change, 9*(6), 472–476. https://doi.org/10.1038/s41558-019-0489-6

Moore, C. (2015). Moral disengagement. *Current Opinion in Psychology, 6,* 199–204. https://doi.org/10.1016/j.copsyc.2015.07.018

Mulvey, K., Shulman, S., Anderson, D., Cole, N., Piepenburg, J., & Sideris, J. (2015). *The climate deception dossiers: Internal fossil fuel industry memos reveal decades of corporate disinformation.* Union of Concerned Scientists. http://www.ucsusa.org/sites/default/files/attach/2015/07/The-Climate-Deception-Dossiers.pdf

Nicolai, S., Franikowski, P., & Stoll-Kleemann, S. (2022). Predicting pro-environmental intention and behavior based on justice sensitivity, moral disengagement, and moral emotions – Results of two quota-sampling surveys. *Frontiers in Psychology, 13.* https://doi.org/10.3389/fpsyg.2022.914366

Ribot, J. (2022). Violent silence: Framing out social causes of climate-related crises. *The Journal of Peasant Studies, 49*(4), 683–712. https://doi.org/10.1080/03066150.2022.2069016

Schäfer, M. (2015). Climate change and the media. *International encyclopedia of the social & behavioral sciences* (pp. 853–859). Elsevier.

Sheeran, P., & Webb, T. L. (2016). The intention-behavior gap. *Social and Personality Psychology Compass, 10*(9), 503–518. https://doi.org/10.1111/spc3.12265

Sligo, F. X., & Jameson, A. M. (2000). The knowledge-behavior gap in use of health information. *Journal of the American Society for Information Science, 51*(9), 858–869. https://doi.org/10.1002/(sici)1097-4571(2000)51:9

Stoll-Kleemann, S., & O'Riordan, T. (2020). Revisiting the psychology of denial concerning low-carbon behaviors: From moral disengagement to generating social change. *Sustainability, 12*(3), 935. https://doi.org/10.3390/su12030935

Stoll-Kleemann, S., O'Riordan, T., & Jaeger, C. (2001). The psychology of denial concerning climate mitigation measures: Evidence from Swiss focus groups. *Global Environmental Change*, *11*(2), 107–117. https://doi.org/10.1016/s0959-3780(00)00061-3

Stoll-Kleemann, S., Franikowski, P., & Nicolai, S. (2023). Development and validation of a scale to assess moral disengagement in high-carbon behavior. *Sustainability*, *15*(3), 2054. https://doi.org/10.3390/su15032054

Supran, G., & Oreskes, N. (2017). Assessing ExxonMobil's climate change communications (1977–2014). *Environmental Research Letters*, *12*(8), 084019. https://doi.org/10.1088/1748-9326/aa815f

Van Zomeren, M., Spears, R., & Leach, C. W. (2010). Experimental evidence for a dual pathway model analysis of coping with climate crisis. *Journal of Environmental Psychology*, *30*, 339–349. https://doi.org/10.1016/j.jenvp.2010.02.006

Van Zomeren, M., Saguy, T., & Schellhaas, F. M. H. (2013). Believing in "making a difference" to collective efforts: Participative efficacy beliefs as a unique predictor of collective action. *Group Processes and Intergroup Relations*, *16*, 618–634. https://doi.org/10.1177/1368430212467476

Walasek, L., & Brown, G. D. (2016). Income inequality, income, and internet searches for status goods: A cross-national study of the association between inequality and well-being. *Social Indicators Research*, *129*(3), 1001–1014. https://doi.org/10.1007/s11205-015-1158-4

Welzer, H. (2011). *Mental infrastructures: How growth entered the world and our souls*. Heinrich Böll Foundation.

World Economic Forum. (2024, January 10). *The global risks report 2024*. https://www.weforum.org/publications/global-risks-report-2024/

Wullenkord, M., & Hamann, K. R. S. (2021). We need to change: Integrating psychological perspectives into the multilevel perspective on socio-ecological transformations. *Frontiers in Psychology*, *12*. https://doi.org/10.3389/fpsyg.2021.655352

5 Toward Climate-Just Behavior: Addressing and Overcoming the Identified Barriers

Susanne Stoll-Kleemann

Actions speak louder than words.

(paraphrased after Johann Wolfgang von Goethe)

Introducing Multiple Complex Approaches toward a More Climate-Just Society

It is evident that proposing effective transformative strategies to mitigate the climate crisis is a highly complex global issue and requires a strategic integration of multiple complimentary approaches (Meadows, 2008; O'Connor et al., 2021; Rau et al., 2024). Therefore, the often proposed solution to distribute knowledge on the impacts of high-carbon behavior is only one piece in the jigsaw puzzle (Bolderdijk et al., 2013; Gifford, 2011; Stoll-Kleemann, 2019). Some colleagues actually emphasize that "moral and educational approaches have generally disappointing track records, and even incentive- and community-based approaches rarely produce much change on their own" (Stern, 2000, p. 419f.). On the contrary, the most efficient behavior modification approaches entail utilizing various sorts of interventions and have to be differentiated according to the kind of behavior that should be modified (e.g. low or high-cost behavior, see Rau et al., 2024). These findings emphasize the constraints of explanations that rely just on one variable when attempting to guide behavior change initiatives. As we have seen in Chapter 3, behavior is influenced by several variables, which often interact with each other (Eyster et al., 2022; Roth, 2016; Sapolsky, 2017; Stern, 2000; Stoll-Kleemann, 2019). The interventions we suggest respect these insights and comprise approaches that consider several factors to achieve a more climate-just society (see Table 5.1).

Approaches Directed to Personal and Social Factors

Appropriate knowledge, values, and attitudes are necessary conditions, but in themselves they are not sufficient to foster changes in climate-unjust behavior (Dibb & Fitzpatrick, 2014; Stoll-Kleemann, 2019; Verbeke, 2008). Of course,

DOI: 10.4324/9781003033547-5

Table 5.1 Table summarizing factors, barriers, and approaches directed to climate-just behavior.

Factors (Internal)	Barriers	Approaches
Knowledge, personality traits, self-efficacy	• Lack of knowledge of the consequences of high-carbon behavior • Lack of skills (*action knowledge*) related to conducting low-carbon behavior • Denial mechanisms provided by moral disengagement block new knowledge • Lack of conscientiousness (as an important personality trait) • Low perceived self-efficacy	• Campaigns and storytelling based on emotional and symbolic messages, e.g. using art, literature, and film making the climate crisis more relatable and urgent • Connecting actions to consequences: help individuals understand how their actions contribute to the climate crisis • Using social marketing mechanisms • Adapted media reporting • Increasing skills that facilitate low-carbon behavior • Facilitating self-efficacy by different mechanisms such as giving feedback and role models showing low-carbon behavior
Values, attitudes, moral development	• Priority of values that favor high-carbon behavior, e.g. hedonic values and self-centeredness • Moral disengagement blocks adequate values through denial and defense mechanisms • Low level of moral development	• Role models and diffusion via connectors • Visualizing the pain of the victims, e.g. via arts, storytelling and media • Promotion of new social norms
Motivation, emotions, psychological needs	• Climate-just behavior seems to be associated with sacrifice and loss of autonomy • Sometimes contradicting needs and emotions at work	• Campaigns and storytelling based on emotional and symbolic messages • Emphasizing how psychological needs and pleasure promotion can be satisfied by low-carbon behavior

(Continued)

Table 5.1 (Continued)

Factors (Internal)	Barriers	Approaches
Habits, fast thinking	• Unconscious, rapid cognitive processing prioritizes existing knowledge • Day-to-day habits as unconscious routine • All kind of different heuristics and biases limit conscious decisions toward low-carbon behavior	• Infrastructure supportive of low-carbon behavior (public transport, etc.) • Support the establishment of new habits, e.g. by using the window of opportunity of a life course transition or disruption
Justification: cognitive dissonance and moral disengagement	• Denial mechanisms provided by cognitive dissonance and social norms which block the incorporation of ethical attitudes into behavior, also on the collective level	• Counter-arguments based on facts and logic in constructive dialogue sessions • Empathetic listening to understand the reasons behind others' moral disengagement and address them respectfully • Self-reflection excises

Factors (external)	Barriers	Approaches
Socio-cultural factors including religion Social identity, social norms Sociodemographic factors	• Symbolism attached to high-carbon behavior • Cultural belief that meat provides strength and vigor (in particular to men) • High income is a predictor for high-carbon behavior as a social marker in the construction of social identities and lifestyles (e.g. as a sign of prosperity)	• Promotion of new cultural norms • Sufficiency as a new lifestyle • Focus on social relations and happiness instead of material wealth • Enhancing social status of an overall responsible low-carbon lifestyle
Politico-economic factors	• Focus on growth and profit maximization • Lack of political will mainly due to powerful lobbies in all related industries • Detrimental subsidies and misleading prices of harmful products and activities (such as air travel) also due to failure of internalizing external costs	• Limiting lobbying efforts • Measures toward more equality such as wealth and inheritance tax • Financial incentives: true prices and taxes on harmful production • Eliminating detrimental subsidies

(*Continued*)

Table 5.1 (Continued)

Factors (Internal)	Barriers	Approaches
Media influence	• Media coverage of the climate crisis is insufficient • Lack of independent media that is not sufficiently critical toward the failure of governmental climate politics and detrimental business interests	• Suitable and pertinent media coverage • Establish independent media platforms • Democratization of information dissemination • More real investigative journalism critical toward lobbies which block climate policies
Availability of and access to a low-carbon infrastructure	• Lack of infrastructure that facilitates low-carbon behavior such as sufficient and affordable public transport; lack of vegetarian-friendly shopping and dining environments • Dependence on fossil fuels	• Increase in affordable sustainable products and public transport • Decrease of high-carbon infrastructure (e.g. parking) • Extension of renewable energies

it is important to directly address common rationalizations in relation to moral disengagement with facts and logic such as to the typical and often used statement *my actions won't make a difference*. For instance, creating platforms for constructively and openly discussing and challenging viewpoints on the climate crisis that contribute to moral disengagement without fear and judgment may be a way forward.

In general, strategies addressing people with high-carbon footprints need to contain ideas that adopting low-carbon behavior will probably be based on the anticipation to take advantage of rewards (or avoid penalty) such as pleasure, financial gain, and an increase in reputation at a certain point of time in a given situation. Examples could be to highlight the personal and immediate advantages of low-carbon behaviors, such as improved health from walking or cycling, savings from reduced energy consumption, or the enjoyment of local, fresh produce. In this context, it is also important to tailor the message to resonate with individual or cultural values. For some, this might mean emphasizing the health benefits for their family or it could be the economic savings. For others, the pleasure component is most important and it is necessary, for example, to gamify sustainable actions via challenges, competitions or apps that track and reward low-carbon behaviors which make the process enjoyable and engaging. However, communicating how already even small, individual actions explained in a specific and achievable manner can make a difference when collectively adopted is crucial. This counters feelings of helplessness, emphasizes agency and empowerment, and reduces the complexity and perceived difficulty of engaging in low-carbon behaviors.

Generally speaking – naturally given that basic needs are fulfilled anyway – each approach directed to achieve low-carbon behavior should consider how one or more of the six psychological needs introduced in Chapter 3 (and outlined by Eyster et al., 2022) can be addressed, namely (1) pleasure promotion, (2) pain prevention, (3) consonance, (4) competence, (5) relatedness and (6) autonomy. Addressing these psychological needs is crucial for mental health and personal development and it is probable that when they are met, individuals are more likely to experience higher levels of well-being and are less likely to experience psychological distress and finally less moral disengagement in the face of the necessity of changing behavior toward more climate justice. To reduce moral disengagement, it is also important to practice empathetic listening to understand the reasons behind others' moral disengagement and address them respectfully. Ideally, this could lead to the encouragement of practices that enable individuals to reflect on their values and actions, such as mindfulness or journaling, which could address the psychological need of consonance.

These needs also play an important role (the last three form the core of self-determination theory of Ryan and Deci (2000) introduced in Chapter 3) when the appropriate incentives for approaches toward less carbon behavior have to be selected. In addition, the insights of intrinsic and extrinsic motivation should be considered. Intrinsic motivation refers to engaging in an activity for its own sake because it is inherently enjoyable or satisfying. When individuals are intrinsically motivated, they are more likely to initiate behavior change because they find the activity personally meaningful, interesting, or aligned with their values (Ryan & Deci, 2000). Knowledge about what intrinsically motivates individuals toward low-carbon behavior can help in designing interventions or behavior change strategies that tap into their inherent drives and passions, making the change process more engaging and sustainable. Examples could be to become engaged in a group toward political climate action or develop a climate-friendly start-up.

Extrinsic motivation involves engaging in an activity to attain some separable outcome or reward, such as praise, money, or recognition. While extrinsic motivators can be effective in initiating behavior change as well, they may not sustain it over the long term if the external rewards are removed (Ryan & Deci, 2000). However, understanding the specific extrinsic motivators that resonate with individuals can still be useful in kick-starting behavior change efforts such as saving money due to energy-saving behavior. For example, offering incentives or rewards that are meaningful to individuals can help them overcome initial barriers and take the first steps toward change. Awards, public recognition, or even simple thank-you messages can be powerful motivators, not only to individuals but also to communities or companies. The most effective form of extrinsic motivation is one that supports individuals' sense of autonomy and competence, for example, to consume and cook in a low-carbon way. This means providing choices, acknowledging their perspectives, and fostering a sense of mastery over the behavior change process. By integrating extrinsic motivators in a way that respects individuals' autonomy, the initiation of behavior change can be smoother and more sustainable.

Ryan and Deci (2000) emphasize the importance of creating environments that support individuals' basic psychological needs for autonomy, competence, and relatedness. Knowledge about these needs can inform the design of environments, interventions, and social support systems that facilitate behavior change initiation. For example, providing informational support that enhances individuals' competence, or creating a supportive social network that fosters relatedness, can encourage them to take the first steps toward change (Ryan & Deci, 2000).

In any case, emotions in a wider sense considering psychological needs and based on the motivational processes outlined above have a stronger influence on behavioral changes than providing knowledge or focusing on attitudes and values (Allen, 2015; Kollmuss & Agyeman, 2002; Piazza et al., 2015). Although, consequently, ways to promote low-carbon behavior might include the provision of information about alternatives to high-carbon behavior, it makes no sense to "merely make rational appeals to people to change behavior based on factual and logical arguments" (Darnton & Evans, 2013, p. 13). Instead, it appears more useful to "provide emotional and empathetic messaging" tailored to the needs of specific groups (Darnton & Evans, 2013, p. 13). Feelings regarding responsibility, guilt, and pride are particularly important because they are strongly connected to the willingness to make sacrifices and change behavior (Antonetti & Maklan, 2014; Bamberg & Möser, 2007; Han et al., 2016; Jefferson et al., 2015; Markowitz & Shariff, 2012).

Knowing that moral and emotional processes influence individuals to reduce their carbon emissions it is a useful strategy to initiate these processes via adequate media coverage. In our study – which is summarized in Stoll-Kleemann et al. (2022) – we successfully demonstrated that exposure to information regarding the victims of climate change elicited such moral and emotional responses as our study participants expressed their intention to modify their behavior if they were regularly exposed to such media coverage. Such a strategy needs to consider that it is necessary that the exposure to these emotional experiences needs to be repeated to generalize emotions, especially those that involve self-condemnation and other-suffering emotions for others. The participants in our study exhibited other-suffering emotions in response to news content and expressed a high likelihood of changing their personal behavior in the face of increased reporting on the climate crisis. It is vital to note that both the frequency and content of media reporting, particularly when it is connected to moral emotions, is important to consider in a strategy which should motivate individuals to modify their behavior. It is also crucial to highlight a scientific consensus to enhance understanding of the origins and consequences of the climate crisis, as well as the role that individuals play in it. More specifically, the media should make the link between one's own advantages and high carbon harmful actions, as well as the disadvantages faced by others, more evident. In such a strategy not solely *adverse* emotions like guilt can function as a moral impetus for doing actions also self-praising emotions result in a propensity to take actions that benefit oneself (Landmann, 2020). For instance, the media could present examples of low carbon-emitting behavior because they also provide a means to experience *positive* emotions like pride.

Given the effectiveness of social influence of role models in driving behavioral change, a viable method would involve increasing the visibility of opinion leaders, who can serve as exemplary examples for such behavior. The *conventional* media (such as the TV) still play an important role because they "take in stories and attitudes from other people and transmit them as social norms to a huge audience" (Cooney, 2011, p. 166). For example, long-running serial dramas can serve as principal vehicles for promoting personal and social changes because "by dramatizing alternative behaviors and their effects on the characters' lives, the dramas help people make informed choices in their own lives. [...] Story lines that dramatize viewers' everyday lives and functional solutions get them deeply involved. Unlike brief exposures to media presentations that typically leave most viewers untouched, ongoing engagement in the evolving lives of models provides numerous opportunities to learn from them" (Bandura, 2016, p. 419f.). Positive role models – not only in the media but everywhere in the surrounding also real world – are also useful to enhance self-efficacy which is an important factor to achieve low-carbon behavior as explained in Chapter 3. Bandura (1997) posits that seeing people similar to oneself succeed by sustained effort raises observers' beliefs that they also possess the capabilities to master comparable activities to succeed. Vicarious experiences involve observing other people successfully completing a task. When one has positive role models in their life one is more likely to absorb at least a few of those positive beliefs about the self. In addition, when done in the right manner, concise and frequent feedback can be one of the most important sources to deepen the self-efficacy achieved by replicating role models which show low-carbon behavior.

In general, making role models more visible is a proven strategy because most people do not decide which behaviors to choose or which attitudes to hold based on a careful analysis (Cooney, 2011). Instead, people change their behavior through "the power of social modeling" and use the available information for their self-development (Bandura, 2016, p. 416). This is supported by Higgs (2015), who concludes that "humans have a highly developed capacity to learn from the behavior of others and to find the approval of others awarding and disapproval aversive" (p. 38). This is why approaches based on reputation and self-image of individual actors – reflecting larger social norms – work so well (Bamberg & Möser, 2007; Barth et al., 2016; Lubchenco et al., 2016).

We know from Chapter 3 that social and cultural norms are "potent and pervasive" and as such are strong and closely interrelated barriers since they function as an excuse for or even legitimization of all kinds of high-carbon behavior while at the same time helping to intimidate people who depart from this accepted behavior and fear social disapproval (Higgs, 2015, p. 42). Therefore, moral disengagement and the difficulty related to change habits and high-carbon behavior may be lessened by the promotion of new social norms, for example, by encouraging people to move in widening social circles that show responsible low-carbon behavior (O'Riordan & Stoll-Kleemann, 2015). This can be achieved by means of different strategies. One is to stress the role of opinion leaders as role models for those who feel insecure about their decision to conduct low-carbon behavior or feel under social pressure. Role models could be a way to enable participants to feel pride and personal esteem in *doing the sustainably right thing* even when others around them are not doing the same. This could help to neutralize the powerful effect of social pressure.

One additional approach in support of this process can be facilitated by techniques used in community-based social marketing (Barr et al., 2011; Stoll-Kleemann & Schmidt, 2017). Findings from neuroscience also support the usefulness of social marketing strategies to influence our climate-related behaviors, for example, because they are able to limit the Consumption-Happiness Myth (Brannigan, 2011). Although social marketing "takes a page from the playbook of traditional advertising" to create behavior change (Cooney, 2011, p. 171), it is rather based on the idea that norms, commitment, and social diffusion have at their core the interactions of individuals in a community and aim at developing supportive social interaction (McKenzie-Mohr, 2000). Similarly, Noppers et al. (2014) found that "the more people think that adopting a sustainable innovation has positive outcomes for their self-identity and social status, the more likely they are to adopt sustainable innovations" (Noppers et al., 2014, p. 60). Because some people see low-carbon behaviors as this type of sustainable innovations, for example, driving an electric car or using solar panels, this seems to be a very promising approach to highlighting the exciting progress and developments in green technology and sustainable practices.

The authors recommend that "targeting symbolic attributes might need subtle and indirect methods as well" and employing lessons that "can possibly be drawn from promotion strategies of high status and innovative brands" (Noppers et al., 2014, p. 61). As long as an advertisement ban for unsustainable products and behavior is not realistic, it makes sense to use the insights and efficient methods of marketing to initiate low-carbon behavior or buy sustainable long-living products if it is unavoidable to buy them or even better to advertise a minimalistic lifestyle.

If cultural and social norms shift, so do the external settings and infrastructure that influence low-carbon behavior. For instance, if we take meat-eating behavior as an example it is easier to eat differently if there are an increasing number of high-quality vegetarian restaurants or vegan outlets nearby. Loyalty and conformity to social reference groups and role models help to determine values, interpretations and emotions. If the core beliefs of the reference group shift and its dominant behavior is opened up to refreshing reinterpretations, then *new habituations* can develop (see below; O'Riordan & Stoll-Kleemann, 2015). If it was the *norm* to consume sustainable products only, and not to fly or use plastics, or even to consume less and more sustainably in general, habits could be readjusted and become embedded in social practices and form a *new conformity*. Realizing that small reductions remain very important in a collective behavior sphere – coupled with a strong sense of *starting together* and establishing a collective efficacy instead of waiting for others to act first – will eliminate the fear of individual sacrifices. In this way, feelings of a new social identity and more accommodative lifestyles may begin to appear.

An approach based on increasing reputation and toward a positive self-image can create conditions that also incentivize companies and countries, and not only individuals, to engage in activities that support sustainability (Lubchenco et al., 2016). They also highlight that altruism, ethical values and reciprocity can be powerful drivers of change because the intrinsic desire of individuals for a positive self-image or to be seen by others in a certain positive way prompts individual actors to do good to achieve personal satisfaction and finally consonance, as one of the six psychological needs mentioned above and in Chapter 3 (see also Griskevicius et al., 2010). "This type of motivation can also apply when groups of actors work together to achieve a goal, creating a sense of camaraderie and shared investment that drives behavior. Even the perception of collective behavior can act as an incentive" (Lubchenco et al., 2016, p. 14511; see also Barth et al., 2016).

However, it is a useful strategy to focus on a small number of people – innovators – which are willing to try out new ideas and behaviors. Of course, new ideas and behaviors that are more *fit* than older ones can radiate, as a growing number of people gradually adopt them. If conditions are right, these *fitter* ideas gradually replace older beliefs for a substantial portion of society (Christakis & Fowler, 2009; Cooney, 2011). In the case of low-carbon behavior, it is necessary that the perception of *losing* something (such as materialistic goods or comfort) needs to be reversed and transferred into a perception of gaining a *good life* relieved from unnecessary ubiquitous consumption and *fulfillment through non-consumer experiences* such as a re-connect to the pleasures of social life and feeling nature (Amel et al., 2009).

Focusing on these innovators and early adopters can help build up the number of supporters for low-carbon behavior until it reaches the critical growth stage. At this point, the power of social networks kicks in, and the majority of the public begins to accept these ideas and behaviors due to having heard about them from friends and neighbors and having observed them in these people's own behavior (Barth et al., 2016; Christakis & Fowler, 2009). Although research on social networks demonstrates that whenever we convince one person to make a change, it "will likely lead others to make a change, and we are more successful than we think", some people – namely, opinion leaders and *connectors* – are linked to and can reach out to many more people than others; they are therefore much more influential than the average person (Cooney, 2011, p. 152f.; Christakis & Fowler, 2009). Connectors are people who have a large number of contacts across an array of social, cultural, professional and economic circles and make a habit of introducing people who work or live in different circles to each other (Christakis & Fowler, 2009; Gladwell, 2000). Opinion leaders such as politicians, prominent business people or entertainers, and religious and civic leaders are also directly linked to a large number of people, although their main impact is in transmitting social norms through the culture, public policy decisions and social media (such as Twitter), with the latter ones gaining increasing importance, including in the area of climate justice. Convincing opinion leaders to support the idea of low-carbon behavior is a critical step in the diffusion process.

Changing Habits and Limiting Fast Thinking

From neuroscience (see Chapter 3), we have learned that to increase pro-environmental behaviors it is necessary to stimulate the brain regions involved in reward integration while we have to avoid decreasing environmentally harmful behaviors (e.g., lowering the heating), they showed increased activity in regions involved in loss anticipation and cognitive control (Brevers et al., 2021). Increasing pro-environmental behaviors is more feasible than decreasing their environmentally harmful behaviors (Doell et al., 2021). Doell et al. (2023) emphasize that "this dissociation at the neural level may help to better understand why people are able to adopt new pro-environmental behaviors while simultaneously continuing to persist with environmentally harmful habits" (Doell et al., 2023, p. 1288). In general, "the importance of habits in sustainable behavior and the consequences for actors such as companies or public authorities attempting to shape the behavior of citizens cannot be overestimated. The stronger the habit, the harder it is to convince citizens (by means of commercials, marketing activities, awareness campaigns, and so on) to change behavioral patterns, if these are not in line with the objectives envisaged by the agents" (Lanzini, 2018, p. 40).

The reasons are that habits are formed through repeated behaviors, which create strong neurological pathways in the brain. These pathways become ingrained over time. Habits often provide some form of reward such as pleasure or comfort with the consequence that the brain's reward system reinforces these behaviors so that they become automatic over time, requiring minimal conscious effort. Habits are often triggered by specific cues or environmental factors, such as time of day or location, and are also shaped by social and cultural norms. Some habits are deeply intertwined with our emotions, serving as coping mechanisms for stress, anxiety and others (Kahneman, 2011;

Duhigg, 2012). Breaking these habits may require addressing underlying emotional issues and finding alternative ways to cope. It can help to demonstrate how low-carbon options can often be more convenient, such as avoiding traffic by cycling, enjoying train rides by reading or talking to people or saving time with efficient home appliances.

A way out of the so-called trap is offered by the habit discontinuity hypothesis, proposed by Verplanken and Wood (2006). It suggests that life transitions and disruptions can create windows of opportunity for behavior change by disrupting existing habits. According to the habit discontinuity hypothesis, major life transitions – such as moving to a new city, starting a new job, or experiencing a significant personal change – can disrupt established habits. During these periods of disruption, individuals may be more open to adopting new behaviors or breaking old habits. This is because the cues and contexts that previously triggered the habitual behavior may no longer be present or as salient. Lanzini (2018) describes the case of commuters for whom a temporary closure of a highway could be the window of opportunity to choose a more sustainable alternative because they may find themselves off-the-hook of old habits and consequently continue with the new more sustainable habit. The results of Verplanken and Roy show that "behaviour change interventions may thus be more effective when delivered in the context of major habit disruptions" (Verplanken & Roy, 2016, p. 128).

Similar to habits, the concept of *fast thinking* is associated with cognitive processes that involve quick, automatic and intuitive decision-making (Kahneman, 2011). It relies on heuristics and mental shortcuts, making it susceptible to biases and errors, which is – as outlined in Chapter 3 – a major barrier for low-carbon behavior. Among them are hyperbolic discounting (prioritizing immediate rewards over future benefits or consequences), status quo bias, and loss aversion, optimism bias or confirmation bias. Understanding and addressing these cognitive biases and heuristics is crucial for designing effective environmental policies and interventions to encourage low-carbon behaviors. Solutions include simplifying sustainable low-carbon choices and effectively communicating the immediate benefits of these choices, for example, by framing information about the climate crisis and its consequences in a way that resonates with people's immediate concerns and values, for example, emphasizing the local, direct impacts of the climate crisis or making low-carbon options the default choice where possible. For example, automatically enrolling residents in renewable energy programs (with the option to opt out) can lead to higher participation rates than requiring individuals to opt. A further way is behavioral nudges, for example, placing vegetarian options more prominently on a menu to encourage lower meat consumption or setting eco-friendly options as default in appliances and vehicles. By addressing cognitive biases and heuristics head-on and creating an environment where sustainable choices are easier, more visible, and more socially supported, it is possible to shift behavior toward lower carbon emissions on both individual and collective levels.

Politico-Economic Factors

Our current growth-driven capitalistic system is a significant contributor to climate injustice and other environmental and social problems limits (Chomsky, 2004; Felber, 2015; Jackson, 2016; Kemfert, 2010; Klein, 2014; Piketty, 2022; Saito, 2024). It relies on excessive consumerism created by manipulative advertising resulting in demand for *unnecessary* products and backed up by the short-term alleviation of negative emotions but often leading to health problems and shopping addictions (Felber, 2018). Consequently, banning advertisements for high-carbon products (like gas-guzzling vehicles, frequent air travel or energy-intensive goods) would reduce demand for these products and activities and thereby leading to lower carbon emissions (Jackson, 2016). In addition, limiting advertising for high-carbon products might indirectly pressure companies to innovate and develop more sustainable products and services. Advertising not only promotes products but also lifestyles. By restricting advertisements that glorify high-carbon lifestyles, it could contribute to a cultural shift toward valuing climate justice and environmentally responsible behaviors (Felber, 2015). Because completely banning advertisement may be difficult to implement in the short run, at least including information about the climate impact of products or incentivizing the promotion of low-carbon products and services should be considered.

Other core principles of capitalism such as profit maximization, short-termism and systemic inequality also strongly contribute to the climate crisis and climate injustice (Chomsky, 2004; Dolderer et al., 2021; Jackson, 2016; Piketty, 2022; Saito, 2024). Therefore, the question would actually be if these problems could be addressed adequately within a capitalistic system or if we need a transition toward a completely different economic system such as postgrowth or degrowth (e.g. Jackson, 2016; Kallis et al., 2015), or common good economics (Felber, 2018), or even participatory socialism (Piketty, 2022), and degrowth communism (Saito, 2024). As all four are similar and we have limited space here we concentrate on the first two ones.

Degrowth is a political, economic and social movement advocating for the downscaling of production, consumption and social inequality (e.g. Jackson, 2016; Kallis et al., 2015). It directly addresses overconsumption, resource depletion and environmental degradation, while it focuses on well-being and quality of life rather than traditional economic growth metrics. It argues that continuous economic growth is unsustainable in a finite world, especially considering environmental crises like climate change. Degrowth promotes reducing the overall material and energy throughput in the economy, leading to lower carbon emissions (Jackson, 2016; Kallis et al., 2015). Encouraging a lifestyle focused on essential needs rather than consumerist desires naturally reduces the carbon footprint associated with production and consumption. It proposes minimizing waste through recycling, reusing and repairing, which is less carbon-intensive than producing new goods as well as the efficient use of resources and extending the life of products reduces the overall carbon footprint (Jackson, 2016). In summary, it presents a radical shift from current consumption and production patterns, focusing on reducing the ecological footprint and fostering a more harmonious relationship with the environment.

From its basic theoretical idea, common good economics is similar to the degrowth concept because it also criticizes massively global economic growth with its negative impacts such as the massive degradation of global ecosystems and extreme inequality (Anthem, 2016; Felber, 2018; Halbmeier & Grabka, 2019; WBGU, 2014). The difference is that common good economics is a more practical economic strategy which meanwhile is considered a viable approach for sustainable transformation across Europe, and even partially worldwide (Dolderer et al., 2021). For example, on their website they name 148 communities in Europe which are involved and even more than 1,000 companies (ECOnGOOD, n.d.a).

The theoretical background of this approach draws on interdisciplinary insights from ethics, ecology, political science, social psychology, doughnut economics or sustainable economics (Felber, 2015; Rogall, 2015; ECOnGOOD, n.d.b; Costanza et al., 2015; Raworth, 2017). The economy for the common good is explicitly committed to the goal of satisfying needs from a long-term societal and ecological perspective. Conceptually, it follows a long historical tradition that the economy as a whole must serve the common good (Dierksmeier, 2016; Dolderer et al., 2021). The definition of *common good economics* is "the science of the satisfaction of the needs of living and future human generations, in alignment with democratic values and ecological planetary boundaries" (Felber, 2019, p. 258). Common good economics emphasizes that human needs can also be satisfied beyond markets, that is, in households, neighborhoods, through commons and public goods (Helfrich & Bollier, 2019). Furthermore, essential resources are not necessarily organized through market transactions: from successful relationships to public security and ecosystem services (Goodwin et al., 2014).

These approaches together with Piketty (2022), Stiglitz (2013), Raworth (2017), Chomsky (2004), Klein (2014), and others argue for the redistribution of wealth and power to limit systemic inequality which is one of the most severe reasons for the climate crisis, for example, because richer individuals and nations, who consume and pollute much more have a very disproportionate impact on the climate (for more explanations, see Chapter 3). Consequently, downsizing inequality would help decreasing greenhouse gas emissions. There are several comprehensive proposals how this can be achieved, for example, with several economic reforms also referring to important institutions such as the World Trade Organization, the global finance system in particular referring to banks that are too big to fail, etc. Furthermore, a comprehensive tax reform would make sense. As this would exceed the scope of a chapter on climate-just behavior, we pick out two cores but singular measures that are important for achieving such a redistribution, namely a wealth tax as globally implemented as possible and the 20–10 justice formula for income limits. Of course, a much higher inheritance tax and/or a limit to inheritance is also a very promising approach. Wealth taxes can complement existing tax systems by making them more progressive. Unlike income taxes, which primarily tax earned income, wealth taxes target accumulated assets and net worth which is important. By incorporating wealth taxes into the tax system, governments can ensure that the wealthiest individuals contribute at least a fair share to society's collective needs (e.g. Piketty, 2022; Stiglitz, 2013).

Nevertheless, there are also several ideas how income injustice can be reduced. One of them is the formula for justice 20–10 by Felber (2006). Because achieving complete equality is not possible, Felber (2006) argues that we have to set limits to inequality and proposes that top incomes should not be more than 20 times the minimum wage, and no one should be allowed to accumulate more than 10 million euros in private wealth. He calls this the *20–10 rule* which would not be an expropriation of the current elite, but the limitation of their right to appropriation, which is only granted to them by society anyway (Felber, 2006). The basic mechanism of the market would remain intact but would be downgraded from an unconditional end in itself to an instrument, placed somewhat more in the service of the common good. 20–10 would finally force people not to satisfy all their needs with money, as they would have to score points with social competence and emotional intelligence or other advantages above the agreed limit. All timeless schools of thought emphasize a life that is moderate in material and rich in immaterial values, including feelings, relationships, values, community, spirituality and nature (Felber, 2006; Jackson, 2016).

Interestingly, there are initiatives by the wealthy themselves, for example, the Taxmenow network, which focuses on initiating changes in the tax system and thus structural change for the system as a whole (Taxmenow, 2022). They are an initiative of wealthy people actively working for tax justice in Germany, Austria and Switzerland. They are convinced that wealth inequality as it exists today undermines democracy and harms society. In addition, there are several countries already having implemented wealth taxes since a longer period such as Norway, Switzerland, Spain and France.

Overall, these measures should be implemented as part of a comprehensive strategy that addresses the root causes of inequality and considers the unique socioeconomic context of each country. Additionally, political will, public support and collaboration between government, civil society and the private sector are essential for successful implementation.

Limiting the Adverse Effects of Lobbying

In Chapter 3, it has been explained that efforts to influence politics through lobbying by powerful economic interests have led to a prioritization of economic growth and short-term gains over long-term climate mitigation and that this has resulted in policies that favor industry and commerce at the expense of necessary climate action (Bandura, 2007; Billé et al., 2013; Danciu, 2014; Klein, 2014; O'Riordan & Stoll-Kleemann, 2015; Stern, 2000). In addition, effective lobbying has resulted in policy stagnation, where needed reforms in energy, transportation, agriculture, and other key sectors are significantly delayed. One example is the fossil fuel lobby which has been very effective in lobbying governments to maintain subsidies for fossil fuels, resist or delay regulations on emissions, and support continued exploration and extraction of fossil fuels. This has contributed to the slow response to the climate crisis and the delayed transition to renewable energy sources. Brulle (2018) has shown that the levels of lobbying expenses seem to be associated with the introduction and likelihood of passing substantial climate legislation.

Even worse, there have been numerous cases of astroturfing, the practice of masking the sponsors of an organization to make it appear as though it originates from and is supported by grassroots participants, aimed at downplaying the climate crisis (Lits, 2020). Astroturfing campaigns often seek to create the illusion of widespread public opposition to climate action or sow doubt about the scientific consensus on climate change (Lits, 2020). One of several examples ExxonMobil's funding of climate denial groups. ExxonMobil and other fossil fuel companies have funded various organizations and think tanks. Astroturfing can be used to distort public perceptions of the climate crisis and undermine efforts to address it. It is essential for the public to be aware of these tactics and scrutinize the sources of information they encounter regarding climate change otherwise it can pose a real threat to democracy (Lits, 2020).

The lobbying process has the power to greatly change the control that government decision-makers have over the type and flow of information. This procedure may impede the dissemination of precise scientific information into the decision-making process. There is a lack of public discussion or rebuttal of perspectives presented by professional lobbyists who have confidential meetings with government leaders. This leads to a situation where communication is consistently distorted. This procedure may impede the dissemination of precise scientific information into the decision-making process and poses a threat to democracy by establishing exclusive networks of influential decision-makers that marginalize the general population (Edwards, 2016).

Therefore, to limit these corporate lobbying activities implementing enhanced transparency measures by mandating complete disclosure of lobbying activity and expenditures is necessary. Transparency in lobbying activities, namely by revealing the identities of lobbyists and the objectives they pursue, serves as a deterrent against the exertion of excessive influence. In addition, stringent rules on campaign contributions and expenditures to reform campaign finance need to be established. While arguing for complete prohibitions on specific lobbying practices would be most efficient but difficult to implement, at least mandatory waiting periods for politicians before to engaging in lobbying endeavors are discussed and partially implemented in some countries such as Germany. This can diminish the way of affluent benefactors and businesses in the political procedure, rendering elected officials less subservient to special interests that resist substantial climate action.

Some further proposals to limit lobbying include enhancing conflict of interest regulations to prohibit individuals with a vested interest in fossil fuel businesses and other sectors that contribute to the climate crisis from occupying public positions where they can exert influence over climate policies. More realistic seem establishing independent committees to analyze the environmental consequences of any legislation. These committees should be non-partisan and free from any influence from lobbyists, ensuring unbiased information is used.

Although complete eradication of detrimental lobbying may not be attainable, implementing some of these techniques might effectively mitigate its influence on climate policy. At least, nongovernmental watchdog organizations such as the Climate Investigations Center, the Union of Concerned Scientists, Greenpeace, Lobbycontrol (based in Germany), Transparency International, and investigative media are already doing a good job in overseeing, revealing and combating detrimental lobbying tactics.

Limiting the Adverse Effects of the Wealthy

The broader economic system facilitates a drastic and unhealthy wealth accumulation (and finally inequality), which promotes the high consumption lifestyles of the richest people caught in a vicious circle of consumption and the so-called luxury emissions (Shue, 1993, see Chapter 1) that exacerbate the climate crisis. This wealth often correlates with greater influence over political and economic decisions including control over industries that are major emitters of greenhouse gases, like fossil fuel companies and the ability to influence policies through lobbying and campaign contributions.

Convincing the wealthy to reduce their carbon footprint is a tricky challenge (due to their relation to power) that requires a combination of *subtle* appeals, economic incentives, policy measures and cultural shifts. For example, many of the world's wealthiest individuals and corporations have investments in fossil fuel industries (Dabi et al., 2022). Divestment from these industries and investment in renewable energy and sustainable practices would be crucial steps toward mitigating climate change. Logically, the wealthy have the resources to invest in very diverse climate solutions. This includes not only funding renewable energy projects but also financing climate change research, and climate communication campaigns and interventions, and supporting climate change advocacy groups. The role of philanthropy can be significant in advancing local and global climate action.

Up to now, appeals to a sense of social responsibility and encouraging the view that the wealthy have a moral duty to lead by example in the fight against the climate crisis have not (yet) been very successful. However, if public recognition and praise for those who take significant steps to reduce their luxury emissions would be more visible, that could potentially incentivize others to follow. The same social dynamics as described in detail above for the general public could work as well here: As people often look to their peers for cues on how to act, creating a culture where climate-just behavior is a status symbol can be powerful and promote stories of wealthy individuals and celebrities who are adopting sustainable lifestyles could be very powerful. Finally, and possibly in addition to the former mentioned options, it is necessary to enforce stricter regulations on activities and products that have a high carbon footprint. This includes stricter building codes for energy and water efficiency (including swimming pools and gardens), strict limits on emissions from private jets and yachts and regulations that require companies to reduce their carbon footprint.

Financial Incentives

Financial incentives can also be an important instrument for a climate-just and sustainable transformation. We take one example based on research coming from our research group from the agrifood sector which is central for solving the climate crisis. The social and ecological externalities of agrifood systems including greenhouse gas emissions causing climate injustice are around 10 trillion dollars (FAO, 2023) and reflect the negative consequences of distribution mechanisms (see Chapter 3), environmentally harmful production, and social exploitations (FAO, 2023). Therefore, many stakeholders are urging to make economies more sustainable (Frey & Sabbatino, 2017; IPCC, 2022; FAO, 2023; Ruggeri Laderchi et al., 2024) and the question is how meaningful contribution to reduce externalities and climate injustice can look like.

True prices seem to be an efficient way forward. They explain the externalized costs of (food) production and communicate the subject toward consumers (Michalke et al., 2022). From production to sale, food has social and environmental impacts, for example, through the emission of nitrogen and greenhouse gases or the use of pesticides (Michalke et al., 2023). These effects are not yet reflected in the market price but are borne by society (Pieper et al., 2020). For example, the negative influence of greenhouse gas emissions (e.g. carbon dioxide, methane, nitrogen oxides) on climate change likewise increases the probability of severe weather events that cause infrastructure destruction, crop failures or health problems. Society and especially future generations have to bear these damages or compensate them with tax or insurance money. Therefore, true cost accounting calculates such externalized costs – in this example here – of food production to make them visible within the market price of today's products and aims to internalize them according to the polluter pays principle (Pieper et al., 2020). The sum of the market price and the monetized negative social and ecological consequences of value chains calculated with true cost accounting constitutes the so-called true prices.

Several campaigns and case studies already introduced the concept of true prices and evaluated customers' reactions. Exemplary campaigns are the introduction of an informational second price tag for several plant-based and animal-based products in 2020 in Berlin (Michalke et al., 2022) or the cooperation between a Dutch supermarket and a social enterprise, where products were sold on a voluntary basis for their true price (True Price, 2023). A cooperation between a group of researchers and a German discounter took it one step further and implemented mandatory true prices (only including negative environmental externalities) for nine products in 2023 during a nationwide campaign week where consumers were confronted with the prices also at checkout (Universität Greifswald, 2023). The results show that the product selection featured plant-based and animal-based options as well as differentiated between conventional and organic produce and sales for products being sold for true prices strongly decreased (Universität Greifswald, 2024). This shows how financial incentives or penalization would in fact lead to consumers' behavior change. However, even the acceptance of policy measures targeting the true prices, like information on true prices and additional surcharges or taxes based on true cost accounting decreased throughout the campaign week (Universität Greifswald, 2024). Additionally, it would be difficult to implement true prices without compromising social sustainability as increased food prices hit poorer households disproportionally (Michalke et al., 2022). Therefore, internalizing adverse effects of value chains into the retail price has to be done with care and perhaps be flanked by sociopolitical measures (e.g. Borchert Commission, 2020). Currently discussed approaches at the consumer level include introducing an animal welfare surcharge (Borchert Commission, 2020; Rasidovic et al., 2023) or adjusting VAT rates to reduce the price gap between conventional products and sustainable products as well as cross-finance an agricultural transformation (Oebel et al., 2024).

Overall, it can be shown that true cost accounting pursues the polluter pays principle (OECD, 1975) – the damage-producing individual or organization is responsible for the caused damage – and thus payments would be implemented at the point of origin along the value chain. The primary goal should be preventing negative externalities in the first place, and only working on reducing negative externalities if not otherwise solvable. Therefore, true cost accounting's unique approach of monetizing externalities offers strong potential for the economy and sustainability communication (FAO, 2023; Michalke et al., 2022; Ruggeri Laderchi et al., 2024). Communicating adverse effects of ecologically harmful, unsafe, underpaying and otherwise socially reckless value chains in monetary values displays the price others pay for cheap grocery items (Michalke et al., 2022). This admittedly simple presentation nevertheless places high demands on scientific work. The monetization of complex problems and parameters like biodiversity loss or forced labor is connected to large uncertainties and is likely modeled with major assumptions. However, the benefit of monetizing these costs nonetheless is a common language between the economic, ecological and social sphere and can lead to a transformation toward an economy that speaks the ecological and social truth (Michalke et al., 2022). This can ultimately simplify political and economic decision-making (Michalke et al., 2022; FAO, 2023).

Ultimately, changing diets and other demand-side actions are important toward transforming agrifood systems and achieving climate justice. Regarding GHG emissions around half of the necessary reduction must be achieved by altering consumer behavior (Latva-Hakuni et al., 2023). True cost accounting and true prices could be an important information and guidance option toward climate-just consumption patterns. Regarding other stakeholders of the value chain, it has been repeatedly demonstrated that conventional agriculture and, in particular, of its nitrogen surpluses and the use of mineral fertilizers and pesticides lead to higher environmental costs (Michalke et al., 2023; Pieper et al., 2020). Therefore, it is important to consider recommendations to reorganize the European Union's (EU) Common Agricultural Policy (CAP). The overall volume of the subsidies can immensely promote the ecological, social, and species-appropriate restructuring of agriculture. The distribution mechanism should be changed in favor of appropriate remuneration for ecological, social, and animal welfare-oriented production. In addition, there are proposals to introduce a nitrogen surplus levy (producers), a mineral fertilizer levy (producers) and a pesticide levy (producers) to internalize the environmental impact costs in line with the polluter pays principle and thus reduce the use of these resources (Chakir & Thomas, 2022; Nielsen et al., 2023; FÖS, 2024), which would all underline true cost accounting approaches. The additional revenue could be used to support the agricultural sector in its transformation toward ecological, social and animal welfare-oriented food production.

In addition, the EU is discussing new regulations and wants to introduce the Corporate Sustainability Reporting Directive contemplated by new European Sustainability Reporting Standards which would obligate companies to report transparently and standardized about risks and chances in their value chains. The European Sustainability Reporting Standards even feature elements of true cost accounting as companies must estimate the financial consequences of risks and chances due to the climate crises, biodiversity losses or labor standards in their value chains. This approach shows similarities to abatement costs that are often used in true cost accounting to communicate the risks and benefits of financial actions toward companies.

Situational Context: Providing Infrastructure Facilitating Low-Carbon Behavior

As outlined in Chapter 3, making climate-friendly options more accessible, affordable and convenient facilitates behavior change more effectively. Of particular importance here is infrastructure development and management, which plays a crucial role in facilitating and encouraging low-carbon behavior.

The design, availability and quality of infrastructure directly influence people's lifestyle choices and affect their carbon footprint. The most important measures in this direction would be an efficient, extensive and affordable public transit systems (buses, trains, trams etc.) which can significantly reduce reliance on personal vehicles, leading to lower carbon emissions. One very prominent positive example since a couple of years is the capital city of Denmark, Copenhagen with its well-thought-through traffic network combining public transport with a decent cycling infrastructure also beyond the city border. In Germany, for example, the introduction of the so-called *Deutschlandticket*, which allows anyone to use public transport throughout Germany for 49 euros per month, is a major success.

If individual transport is unavoidable, at least infrastructure for electric vehicles, including charging stations and supportive policies, can accelerate the transition from fossil fuel-powered vehicles to electric ones. In addition, a well-designed, safe cycling and walking infrastructure encourages nonmotorized, zero-emission modes of transportation. Here one positive example is the capital city of France, Paris, which has built up an impressive network of cycling infrastructure in recent years. Paris has also increased its parking fees for SUVs (unnecessarily large cars).

Investments in renewable energy infrastructure (solar, wind, hydroelectric etc.) are already going on but need to be enhanced in a more efficient and rapid way (Kemfert, 2010). Of particular importance here are community-level and decentralized renewable energy solutions because they can reduce transmission losses and empower local low-carbon initiatives. From communities and cities already come some useful urban and rural planning initiatives that promote compact, mixed-use developments that reduce the need for long commutes and lower transportation emissions. In addition, an effective waste management infrastructure can significantly reduce the carbon footprint by minimizing landfill use and associated methane emissions. In addition, urban areas can deliver green infrastructure, including parks, green roofs and permeable pavements, which can also effectively reduce the impact of urban heat islands and control the flow of rainwater. Singapore could serve as an inspiring example of how urban areas can integrate green infrastructure elements, as it has indeed introduced vertical gardens and green roofs, innovative water management, including wastewater recycling and rainwater harvesting, an efficient public transport system and standards and incentives for green building.

For the food sector, an appropriate low-carbon infrastructure would include specific factors of food supply in the environment, such as the type of food, food sources and the availability of and access to food (Dagevos & Voordouw, 2013; Furst et al., 1996; Verain et al., 2015). This sector has seen the most positive development in terms of the availability of low-carbon products: many new products have been developed to complement the classic plant-based diet, such as pulses, tofu, seitan and lupin-based products. The low-carbon *food environment* is growing thanks to an ever-increasing range of tasty and affordable organic, vegetarian and vegan products in supermarkets and on the menus of restaurants, canteens and cafeterias (Stoll-Kleemann & Schmidt, 2017). Food companies and the catering sector already offer organic, regional, meat-free or lower meat alternatives to ready meals to serve the ready meal and out-of-home catering markets. Overall, the development of infrastructure that supports low-carbon behavior could make an important contribution to reducing high-carbon behavior.

Of course, the ideal solution would be achieving circular economy models, where waste is minimized and materials are reused or recycled as much as possible. Finally, robust digital infrastructure supports remote working and virtual meetings, reducing the need for travel and its associated carbon emissions. By focusing on these kinds of sustainable, efficient and innovative infrastructure solutions, it is possible to significantly encourage and facilitate low-carbon behaviors across various sectors of society.

Contributions of Media Reporting to Addressing External Factors

We mentioned above that the media could in principle play a crucial role in addressing the personal and social factors related to the climate crisis with widespread positive outcome, for example, by shaping public awareness and influencing policy debates, regular and repeated reporting on climate-related disasters and emphasizing its severity and the need for immediate action (Stoll-Kleemann et al., 2022). In addition, the media can highlight practical solutions including sustainable practices and stories of successful climate action at individual, community, corporate and governmental levels. Utilizing powerful imagery and compelling narratives can make the abstract and often overwhelming issue of the climate crisis more tangible and relatable to the audience. Unfortunately, our own empirical results show that current media coverage of the climate crisis – at least in Germany – is neither sufficient nor focused enough to induce emotions on the current climate crisis and hence initiate behavioral change (Stoll-Kleemann et al., 2022).

Nevertheless, there are several more possibilities at the systemic level how the media can contribute to help mitigating the climate crisis. Investigative journalism is crucial in holding corporations and governments accountable for their climate-related policies and practices. Exposing negligence, lobbying efforts against climate action, and failure to meet climate commitments can drive policy changes. In addition, actively countering misinformation and climate change denial in the media is crucial. This involves fact-checking, debunking false claims and ensuring that misleading narratives do not go unchallenged (Rahmstorf, 2012). Providing a platform for scientists, environmentalists and policymakers to share expert insights can help in educating the public and policymakers, contributing to a more informed and science-based public discourse. Climate issues can be facilitated by the media through debates, forums and interactive platforms. This can increase public interest and participation in climate-related activities and policies. Including a range of voices, especially those from marginalized and frontline communities most affected by climate change, can provide a more comprehensive view of the impacts of climate change and the need for inclusive and just solutions. In general, it can be observed that, when climate change becomes a prominent topic in the media, it increases the pressure on political leaders to address it.

However, caution must be exercised, as well because message framing as an essential part of motivating communication can be detrimental when economic growth is framed as the default option in messages or advantageous when specific emission-reduction strategies are deemed the most effective to promote engagement. While general doom often results in disengagement to suppress fear, threats to people's immediate surroundings, such as their family, living areas or belongings, can trigger active behavior (Hine et al., 2016; Peeters et al., 2019).

Future Research

It has become clear that it is necessary to conduct more detailed empirical research on effective strategies to increase – in particular rich and famous – people's, companies' and political decision-makers' motivation to act on the climate crisis, and how their transformative actions can be operationalized in daily life. While there is considerable research ongoing in environmental psychology focusing on individual behavior and singular factors on interventions, there is a vast scarcity of research at the level of the external factors. In this respect, the most important future research encompasses a more detailed analysis of how the influence and power of strong corporate high-carbon lobbies on political leaders can be quickly interrupted. Brulle (2018) proposes that an in-depth study examining the effects of legislation on a corporation, its stance on the issue and the level of resources that it allocates would provide valuable insights into the reasons behind corporate lobbying. In addition, he suggests that because there is a scarcity of studies about the influence of lobbying on climate legislation, and thus conducting a comprehensive analysis that considers multiple variables to examine the influence of lobbying would strongly enhance the existing body of research on climate lobbying (Brulle, 2018). The magnitude and scope of lobbying on climate policy necessitate additional scrutiny of this significant political domain. In addition, Lits (2020) suggests that due to the weak rules on lobbying transparency in the EU, it is necessary to "conduct further research on astroturf lobbying in order to understand how astroturf movements are created, how they operate, and how they communicate" (p. 18).

Although we are beginning to understand the motives behind climate-just behavior, further interdisciplinary and integrative research is necessary to learn more about the interconnectedness of factors and the appropriate strategies that it entails. This concerns both levels, the one including the several personal factors as well as the external ones. On a personal level, we can learn a lot from neuroscientific approaches. They provide causal insights into important mechanisms that underlie carbon-related behavior and can ultimately inform the development of more effective interventions (Doell et al., 2023). The findings outlined in Chapter 3 suggest that imagining the plight of future generations may be a more effective way to increase low-carbon behavior. This can be realized with virtual reality techniques; for example, by demonstrating the future consequences of climate change by showing rising sea levels or assessing consumer decisions in more realistic simulated settings (Meijer et al., 2023). These more realistic environments will allow for a more valid evaluation of choices by reducing psychological distance (e.g. Gifford, 2011; Hahnel & Brosch, 2018; Markowitz & Bailenson, 2021). In addition, neurosciences could contribute to find solutions in investigating inter-individual and group differences in a more systematic manner. Doell et al. (2023) suggest that neuroscientific research could focus on the neural markers of climate change-related motivation and analyze how and why specific groups tend to resist adopting sustainable lifestyles and if such groups systematically differ in the ways they process climate change-related information (Doell et al., 2023).

References

Allen, K. (2015). *Emotions & cognitive dissonance: Your supreme divine guidance.* Collective Evolution (CE). http://www.collective-evolution.com/2013/02/05/emotions-and-cognitive-dissonance-theory-your-supreme-divine-guidance/

Amel, E. L., Manning, C. M., & Scott, B. A. (2009). Mindfulness and sustainable behavior: Pondering attention and awareness as means for increasing green behavior. *Ecopsychology, 1*(1), 14–25. https://doi.org/10.1089/eco.2008.0005

Anthem, P. (2016, April 16). *Risk of hunger pandemic as coronavirus set to almost double acute hunger by end of 2020.* World Food Programme (WFP). https://www.wfp.org/stories/risk-hunger-pandemic-coronavirus-set-almost-double-acute-hunger-end-2020

Antonetti, P., & Maklan, S. (2014). Feelings that make a difference: How guilt and pride convince consumers of the effectiveness of sustainable consumption choices. *Journal of Business Ethics, 124*(1), 117–134. https://doi.org/10.1007/s10551-013-1841-9

Bamberg, S., & Möser, G. (2007). Twenty years after Hines, Hungerford, and Tomera: A new meta-analysis of psycho-social determinants of pro-environmental behaviour. *Journal of Environmental Psychology, 27*(1), 14–25. https://doi.org/10.1016/j.jenvp.2006.12.002

Bandura, A. (1997). *Self-efficacy: The exercise of control.* Freeman.

Bandura, A. (2007). Impeding ecological sustainability through selective moral disengagement. *International Journal of Innovation and Sustainable Development, 2*(1), 8–35. https://doi.org/10.1504/ijisd.2007.016056

Bandura, A. (2016). *Moral disengagement: How people do harm and live with themselves.* Worth Publishers.

Barr, S., Gilg, A. W., & Shaw, G. (2011). Citizens, consumers and sustainability: (Re) Framing environmental practice in an age of climate change. *Global Environmental Change, 21*(4), 1224–1233. https://doi.org/10.1016/j.gloenvcha.2011.07.009

Barth, M., Jugert, P., & Fritsche, I. (2016). Still underdetected – Social norms and collective efficacy predict the acceptance of electric vehicles in Germany. *Transportation Research Part F: Traffic Psychology and Behaviour, 37*, 64–77. https://doi.org/10.1016/j.trf.2015.11.011

Billé, R., Kelly, R. P., Biastoch, A., Harrould-Kolieb, E., Herr, D., Joos, F., Kroeker, K. J., Laffoley, D., Oschlies, A., & Gattuso, J. (2013). Taking action against ocean acidification: A review of management and policy options. *Environmental Management, 52*(4), 761–779. https://doi.org/10.1007/s00267-013-0132-7

Bolderdijk, J. W., Gorsira, M., Keizer, K., & Steg, L. (2013). Values determine the (in) effectiveness of informational interventions in promoting pro-environmental behavior. *PLoS One, 8*(12), e83911. https://doi.org/10.1371/journal.pone.0083911

Borchert Commission. (2020). *Empfehlungen des Kompetenznetzwerks Nutztierhaltung.* Borchert Kommission. https://www.bmel.de/SharedDocs/Downloads/DE/_Tiere/Nutztiere/200211-empfehlung-kompetenznetzwerk-nutztierhaltung.pdf?__blob=publicationFile&v=3

Brannigan, F. (2011). Dismantling the consumption-happiness myth: A neuropsychological perspective on the mechanisms that lock us into unsustainable consumption. *Engaging the public with climate change* (pp. 84–99). Earthscan.

Brevers, D., Baeken, C., Maurage, P., Sescousse, G., Vögele, C., & Billieux, J. (2021). Brain mechanisms underlying prospective thinking of sustainable behaviours. *Nature Sustainability, 4*(5), 433–439. https://doi.org/10.1038/s41893-020-00658-3

Brulle, R. J. (2018). The climate lobby: A sectoral analysis of lobbying spending on climate change in the USA, 2000 to 2016. *Climatic Change, 149*(3–4), 289–303. https://doi.org/10.1007/s10584-018-2241-z

Chakir, R., & Thomas, A. (2022). Unintended consequences of environmental policies: The case of set-aside and agricultural intensification. *Environmental Modeling & Assessment, 27*(2), 363–384. https://doi.org/10.1007/s10666-021-09815-0

Chomsky, N. (2004). *Profit over people: Neoliberalismus und globale weltordnung.* Europa Verlag.

Christakis, N. A., & Fowler, J. H. (2009). *Connected: The surprising power of our social networks and how they shape our lives.* Little, Brown Spark.

Cooney, N. (2011). *change of heart: What psychology can teach us about spreading social change.* Lantern Books.

Costanza, R., Cumberland, J. H., Daly, H., Goodland, R., Norgaard, R. B., Kubiszewski, I., & Franco, C. (2015). *An introduction to ecological economics* (2nd ed.). CRC Press.

Dabi, N., Maitland, A., Lawson, M., Stroot, H., Poidatz, A., & Khalfan, A. (2022). *Carbon billionaires: The investment emissions of the world's richest people.* Oxfam GB for Oxfam International. https://doi.org/10.21201/2022.9684

Dagevos, H., & Voordouw, J. (2013). Sustainability and meat consumption: Is reduction realistic? *Sustainability: Science, Practice and Policy, 9*(2), 60–69. https://doi.org/10.1080/15487733.2013.11908115

Danciu, V. (2014). Manipulative marketing: Persuasion and manipulation of the consumer through advertising. *Theoretical and Applied Economics, 2*(591), 19–34. https://doaj.org/article/a128835485c2498fad24213a4462f1e8

104   *Toward Climate-Just Behavior*

Darnton, A., & Evans, D. (2013). *Influencing behaviours. A technical guide to the ISM tool.* The Scottish Government.

Dibb, S., & Fitzpatrick, I. (2014). *Let's talk about meat: Changing dietary behaviour for the 21st century.* Eating Better.

Dierksmeier, D. (2016). *Reframing economic ethics. The philosophical foundations of humanistic management.* Palgrave Macmillan, Springer Nature.

Doell, K. C., Conte, B., & Brosch, T. (2021). Interindividual differences in environmentally relevant positive trait affect impacts sustainable behavior in everyday life. *Scientific Reports, 11*(1), 20423. https://doi.org/10.1038/s41598-021-99438-y

Doell, K. C., Berman, M. G., Bratman, G. N., Knutson, B., Kühn, S., Lamm, C., Pahl, S., Sawe, N., Van Bavel, J. J., White, M. P., & Brosch, T. (2023). Leveraging neuroscience for climate change research. *Nature Climate Change, 13*(12), 1288–1297. https://doi.org/10.1038/s41558-023-01857-4

Dolderer, J., Felber, C., & Teitscheid, P. (2021). From neoclassical economics to common good economics. *Sustainability, 13*(4), 2093. https://doi.org/10.3390/su13042093

Duhigg, C. (2012). *The power of habit: Why we do what we do in life and business.* Random House Business Books.

Edwards, L. (2016). The role of public relations in deliberative systems. *Journal of Communication, 66*(1), 60–81. https://doi.org/10.1111/jcom.12199

Eyster, H. N., Satterfield, T., & Chan, K. M. A. (2022). Why people do what they do: An interdisciplinary synthesis of human action theories. *Annual Review of Environment and Resources, 47*(1), 725–751. https://doi.org/10.1146/annurev-environ-020422-125351

Felber, C. (2006). *50 Vorschläge für eine gerechtere Welt.* Deuticke Verlag.

Felber, C. (2015). *Change everything: Creating an economy for the common good.* Zed Books.

Felber, C. (2018). *Gemeinwohl-Ökonomie.* Pieper.

Felber, C. (2019). *This is not economy: Aufruf zur Revolution der Wirtschaftswissenschaft.* Deuticke Verlag.

Food and Agriculture Organization of the United Nations (FAO). (2023). *The state of food and agriculture 2023. Revealing the true cost of food to transform agrifood systems.* FAO. https://www.fao.org/3/cc7724en/cc7724en.pdf

Forum Ökologisch-Soziale Marktwirtschaft (FÖS). (2024). Subventionen und Abgaben im Agrarsektor. *Welchen Beitrag können sie zu Umweltschutz und Entlastung des Staatshaushalts leisten?.* FÖS. https://foes.de/publikationen/2024/2024-01_Kurzstudie_Subventionen_und_Abgaben_im_Agrarsektor.pdf

Frey, M., & Sabbatino, A. (2017). The role of the private sector in global sustainable development: The UN 2030 agenda. *Palgrave studies in governance, leadership and responsibility* (pp. 187–204). https://doi.org/10.1007/978-3-319-63480-7_10

Furst, T., Connors, M., Bisogni, C. A., Sobal, J., & Falk, L. W. (1996). Food choice: A conceptual model of the process. *Appetite, 26*(3), 247–266. https://doi.org/10.1006/appe.1996.0019

Gifford, R. (2011). The dragons of inaction: Psychological barriers that limit climate change mitigation and adaptation. *American Psychologist, 66*(4), 290–302. https://doi.org/10.1037/a0023566

Gladwell, M. (2000). *The tipping point: How little things can make a big difference.* Little, Brown Spark.

Goodwin, N., Harris, J., Nelson, J. A., Roach, B., & Torras, M. (2014). *Principles of economics in context*. Routledge.

Griskevicius, V., Tybur, J. M., & Van Den Bergh, B. (2010). Going green to be seen: Status, reputation, and conspicuous conservation. *Journal of Personality and Social Psychology, 98*(3), 392–404. https://doi.org/10.1037/a0017346

Hahnel, U. J., & Brosch, T. (2018). Environmental trait affect. *Journal of Environmental Psychology, 59*, 94–106. https://doi.org/10.1016/j.jenvp.2018.08.015

Halbmeier, C., & Grabka, M. M. (2019). *Vermögensungleichheit in Deutschland bleibt trotz deutlich steigender Nettovermögen anhaltendhoch*. *DIW Wochenbericht, 40*, 735–745. https://www.diw.de/documents/publikationen/73/diw_01.c.679972.de/19-40-1.pdf

Han, H., Hwang, J., & Lee, M. J. (2016). The value–belief–emotion–norm model: Investigating customers' eco-friendly behavior. *Journal of Travel & Tourism Marketing, 34*(5), 590–607. https://doi.org/10.1080/10548408.2016.1208790

Helfrich, S., & Bollier, D. (2019). *Frei, Fair und Lebendig – Die Macht der Commons*. Transcript Verlag.

Higgs, S. (2015). Social norms and their influence on eating behaviours. *Appetite, 86*, 38–44. https://doi.org/10.1016/j.appet.2014.10.021

Hine, D. W., Phillips, W. J., Cooksey, R., Reser, J. P., Nunn, P. D., Marks, A. D. G., Loi, N. M., & Watt, S. E. (2016). Preaching to different choirs: How to motivate dismissive, uncommitted, and alarmed audiences to adapt to climate change? *Global Environmental Change, 36*, 1–11. https://doi.org/10.1016/j.gloenvcha.2015.11.002

Intergovernmental Panel on Climate Change (IPCC). (2022). *Climate change 2022: Impacts, adaption and vulnerability – Summary for policymakers*. Intergovernmental Panel on Climate Change. https://www.ipcc.ch/report/ar6/wg2/downloads/report/IPCC_AR6_WGII_SummaryForPolicymakers.pdf

International Federation for the Economy for the Common Good e.V. [ECOnGOOD]. (n.d.a). *What is ECG*. ECOnGOOD. https://www.ecogood.org/what-is-ecg/

International Federation for the Economy for the Common Good e.V. [ECOnGOOD]. (n.d.b). *Home*. ECOnGOOD. https://www.ecogood.org/

Jackson, T. (2016). *Prosperity without growth. Foundations for the economy of tomorrow*. Routledge.

Jefferson, R., McKinley, E., Capstick, S., Fletcher, S., Griffin, H., & Milanese, M. (2015). Understanding audiences: Making public perceptions research matter to marine conservation. *Ocean & Coastal Management, 115*, 61–70. https://doi.org/10.1016/j.ocecoaman.2015.06.014

Kahneman, D. (2011). *Thinking, fast and slow*. Penguin Books Ltd.

Kallis, G., Demaria, F., & d'Alisa, E. (2015). Degrowth. In *Degrowth: A vocabulary for a new era*. Routledge.

Kemfert, C. (2010). Sind wirtschaftliches Wachstum und ein Klimaschutz, der die Erderwärmung bei 2° plus hält, vereinbar? *Magazin der Heinrich-Böll-Foundation, 1*. https://www.boell.de/sites/default/files/BoellThema_1-10_V06_abReader7kommentierbar.pdf

Klein, N. (2014). *This changes everything: Capitalism vs. the climate*. Allen Lane.

Kollmuss, A., & Agyeman, J. (2002). Mind the gap: Why do people act environmentally and what are the barriers to pro-environmental behavior? *Environmental Education Research, 8*(3), 239–260. https://doi.org/10.1080/13504620220145401

Landmann, H. (2020). Emotions in the context of environmental protection: Theoretical considerations concerning emotion types, eliciting processes, and affect generalization. *Umweltpsychologie, 24*(2), 61–73. http://umps.de/php/artikeldetails. php?id=745

Lanzini, P. (2018). *Responsible citizens and sustainable behaviour. New interpretive frameworks.* Routledge.

Latva-Hakuni, E., Bengtsson, M., Coscieme, L., Deventer, M. J., & Wollesen, G. (2023). *Food production and consumption in a 1.5°C world – Options for Germany.* Hot or Cool Institute. https://hotorcool.org/wp-content/uploads/2023/11/Food-Production-and-Consumption-in-a-15-World-final-report-09-21-2023.pdf

Lits, B. (2020). Exploring astroturf lobbying in the EU: The case of responsible energy citizen coalition. *European Policy Analysis, 7*(1), 226–239. https://doi.org/10.1002/epa2.1086

Lubchenco, J., Cerny-Chipman, E. B., Reimer, J. J., & Levin, S. A. (2016). The right incentives enable ocean sustainability successes and provide hope for the future. *Proceedings of the National Academy of Sciences of the United States of America, 113*(51), 14507–14514. https://doi.org/10.1073/pnas.1604982113

Markowitz, D. M., & Bailenson, J. N. (2021). Virtual reality and the psychology of climate change. *Current Opinion in Psychology, 42*, 60–65. https://doi.org/10.1016/j.copsyc.2021.03.009

Markowitz, E. M., & Shariff, A. F. (2012). Climate change and moral judgement. *Nature Climate Change, 2*(4), 243–247. https://doi.org/10.1038/nclimate1378

McKenzie-Mohr, D. (2000). New ways to promote pro-environmental behavior: Promoting sustainable behavior: An introduction to community-based social marketing. *Journal of Social Issues, 56*(3), 543–554. https://doi.org/10.1111/0022-4537.00183

Meadows, D. (2008). *Thinking in systems: A primer.* Earthscan.

Meijer, M. A., Brabers, A., & De Jong, J. (2023). Social context matters: The role of social support and social norms in support for solidarity in healthcare financing. *PLoS One, 18*(9), e0291530. https://doi.org/10.1371/journal.pone.0291530

Michalke, A., Stein, L. C., Fichtner, R., Gaugler, T., & Stoll-Kleemann, S. (2022). True cost accounting in agri-food networks: A German case study on informational campaigning and responsible implementation. *Sustainability Science, 17*(6), 2269–2285. https://doi.org/10.1007/s11625-022-01105-2

Michalke, A., Köhler, S., Meßmann, L., Thorenz, A., Tuma, A., & Gaugler, T. (2023). True cost accounting of organic and conventional food production. *Journal of Cleaner Production, 408*, 137134. https://doi.org/10.1016/j.jclepro.2023.137134

Nielsen, H., Konrad, M. T., Pedersen, A. B., & Gyldenkærne, S. (2023). Ex-post evaluation of the Danish pesticide tax: A novel and effective tax design. *Land Use Policy, 126*, 106549. https://doi.org/10.1016/j.landusepol.2023.106549

Noppers, E. H., Keizer, K., Bolderdijk, J. W., & Steg, L. (2014). The adoption of sustainable innovations: Driven by symbolic and environmental motives. *Global Environmental Change, 25*, 52–62. https://doi.org/10.1016/j.gloenvcha.2014.01.012

O'Connor, M. I., Mori, A., Gonzalez, A., Dee, L. E., Loreau, M., Avolio, M. L., Byrnes, J. E. K., Cheung, W., Cowles, J., Clark, A. T., Hautier, Y., Hector, A., Komatsu, K. J., Newbold, T., Outhwaite, C. L., Reich, P. B., Seabloom, E. W., Williams, L., Wright, A. J., & Isbell, F. (2021). Grand challenges in biodiversity–ecosystem functioning research in the era of science-policy platforms require explicit consideration of feedbacks. *Proceedings of the Royal Society, 288*(1960). https://doi.org/10.1098/rspb.2021.0783

O'Riordan, T., & Stoll-Kleemann, S. (2015). The challenges of changing dietary behavior toward more sustainable consumption. *Environment: Science and Policy for Sustainable Development, 57*(5), 4–13. https://doi.org/10.1080/00139157.2015.1069093

Oebel, B., Stein, L., Michalke, A., Stoll-Kleemann, S. & Gaugler, T. (2024). Towards true prices in food retailing: The value added tax as an instrument transforming agri-food systems. *Sustainability Science.* https://doi.org/10.1007/s11625-024-01477-7

Organisation for Economic Co-operation and Development (OECD). (1975). *The polluter pays principle.* OECD. https://www.oecdilibrary.org/docserver/9789264044845-en.pdf?expires=1635841062&id=id&accname=guest&checksum=3B59AB3B3C545B07DC30BBB865C5BFC6

Peeters, W., Diependaele, L., & Sterckx, S. (2019). Moral disengagement and the motivational gap in climate change. *Ethical Theory and Moral Practice, 22*(2), 425–447. https://doi.org/10.1007/s10677-019-09995-5

Piazza, J., Ruby, M. B., Loughnan, S., Luong, M., Kulik, J., Watkins, H. M., & Seigerman, M. (2015). Rationalizing meat consumption: The 4NS. *Appetite, 91*, 114–128. https://doi.org/10.1016/j.appet.2015.04.011

Pieper, M., Michalke, A., & Gaugler, T. (2020). Calculation of external climate costs for food highlights inadequate pricing of animal products. *Nature Communications, 11*(1), 6117. https://doi.org/10.1038/s41467-020-19474-6

Piketty, T. (2022). *Eine kurze Geschichte der Gleichheit.* C.H. Beck.

Rahmstorf, S. (2012). Is journalism failing on climate? *Environmental Research Letters, 7*(4), 041003. https://doi.org/10.1088/1748-9326/7/4/041003

Rasidovic, A., Oebel, B., Stein, L., Michalke, A., & Gaugler, T. (2023, March 07–10). *Soziale externe Kosten: Ein Framework zur Monetarisierung von Tierwohl zur Berechnung wahrer Lebensmittelpreise.* 16. Wissenschaftstagung Ökologischer Landbau, Frick (CH). https://wissenschaftstagung.de/16-wissenschaftstagung-oekologischer-landbau/

Rau, H., Nicolai, S., Franikowski, P., & Stoll-Kleemann, S. (2024). Distinguishing between low- and high-cost pro-environmental behavior: Empirical evidence from two complementary studies. *Sustainability, 16*, 2206. https://doi.org/10.3390/su16052206

Raworth, K. (2017). *Doughnut economics: Seven ways to think like a 21st-century-economist.* Random House Business Books.

Rogall, H. (2015). *Grundlagen einer Nachhaltigen Wirtschaftslehre* (2nd ed.). Metropolis.

Roth, G. (2016). *Persönlichkeit, Entscheidung und Verhalten. Warum es so schwierig ist, sich und andere zu ändern: Persönlichkeit, Entscheidung und Verhalten.* Klett-Cotta.

Ruggeri Laderchi, C., Lotze-Campen, H., DeClerck, F., Bodirsky, B.L., Collignon, Q., Crawford, M.S., Dietz, S., Fesenfeld, L., Hunecke, C., Leip, D., Lord, S., Lowder, S., Nagenborg, S., Pilditch, T., Popp, A., Wedl, I., Branca, F., Fan, S., Fanzo, J., ..., Songwe, V. (2024). *Global policy report. The economics of the food system transformation.* Food System Economics Commission [FSEC]. https://foodsystemeconomics.org/wp-content/uploads/FSEC-GlobalPolicyReport-February2024.pdf

Ryan, R. M., & Deci, E. L. (2000). Self-determination theory and the facilitation of intrinsic motivation, social development, and well-being. *American Psychologist, 55*(1), 68–78. https://doi.org/10.1037/0003-066x.55.1.68

Saito, K. (2024). *Slow down: How degrowth communism can save the earth.* Orion Publishing Group.

Sapolsky, R. M. (2017). *Behave: The biology of humans at our best and worst.* Vintage/Penguin Random House.

Shue, H. (1993). Subsistence emissions and luxury emissions. *Law & Policy, 15*(1), 39–60. https://doi.org/10.1111/j.1467-9930.1993.tb00093.x

Stern, P. C. (2000). New environmental theories: Toward a coherent theory of environmentally significant behavior. *Journal of Social Issues, 56*(3), 407–424. https://doi.org/10.1111/0022-4537.00175

Stiglitz, J. E. (2013). *The price of inequality: How Today's divided society endangers our future.* W. W. Norton & Company.

Stoll-Kleemann, S. (2019). Feasible options for behavior change toward more effective ocean literacy: A systematic review. *Frontiers in Marine Science, 6*(237). https://doi.org/10.3389/fmars.2019.00273

Stoll-Kleemann, S., & Schmidt, U. (2017). Reducing meat consumption in developed and transition countries to counter climate change and biodiversity loss: A review of influence factors. *Regional Environmental Change, 17*(5), 1261–1277. https://doi.org/10.1007/s10113-016-1057-5

Stoll-Kleemann, S., Nicolai, S., & Franikowski, P. (2022). Exploring the moral challenges of confronting high-carbon-emitting behavior: The role of emotions and media coverage. *Sustainability, 14*(10), 5742. https://doi.org/10.3390/su14105742

Taxmenow. (2022). *Initiative.* Taxmenow. https://www.taxmenow.eu/en/unserearbeit

True Price. (2023). *True pricing experiment Bij Albert Heijn: Factsheet berekeningen/Extern/V2.0.* trueprice.org. https://static.ah.nl/binaries/ah/content/assets/ah-nl/permanent/over-ah/true-pricing-ah.pdf

Universität Greifswald. (2023, July 31). *Wahre Kosten für Lebensmittel werden bei Discounter wissenschaftlich ermittelt* [Pressemitteilung]. https://www.uni-greifswald.de/universitaet/information/aktuelles/detail/n/wahre-kosten-fuer-lebensmittel-werden-bei-discounter-wissenschaftlich-ermittelt-new64c76b78e023b674556727/

Universität Greifswald. (2024, January 23). *Wunsch nach Transparenz von Lebensmitteln ist messbar gewachsen – Ergebnisse der Wahre-Kosten-Kampagne mit Penny liegen vor* [Pressemitteilung]. https://www.uni-greifswald.de/universitaet/information/aktuelles/detail/n/wunsch-nach-transparenz-von-lebensmitteln-ist-messbar-gewachsen-ergebnisse-der-wahre-kosten-kampagne-mit-penny-liegen-vor-new65af6b3d5440e125960425/

Verain, M., Dagevos, H., & Antonides, G. (2015). Flexitarianism: A range of sustainable food styles. *Handbook of research on sustainable consumption* (pp. 209–223). Edward Elgar Publishing.

Verbeke, W. (2008). Impact of communication on consumers' food choices. *Proceedings of the Nutrition Society, 67*(3), 281–288. https://doi.org/10.1017/s0029665108007179

Verplanken, B., & Roy, D. (2016). Empowering interventions to promote sustainable lifestyles: Testing the habit discontinuity hypothesis in a field experiment. *Journal of Environmental Psychology, 45*, 127–134. https://doi.org/10.1016/j.jenvp.2015.11.008

Verplanken, B., & Wood, W. (2006). Interventions to break and create consumer habits. *Journal Of Public Policy & Marketing, 25*(1), 90–103. https://doi.org/10.1509/jppm.25.1.90

Wissenschaftlicher Beirat der Bundesregierung Globale Umweltveränderungen [WBGU]. (2014). *Human progress within planetary guard rails – A contribution to the SDG debate: Policy paper 8*. WBGU. https://www.wbgu.de/fileadmin/user_upload/wbgu/publikationen/politikpapiere/pp8_2014/wbgu_pp8_en.pdf

Index

Note: *Italicized* and **bold** pages refer to figures and tables respectively, and page numbers followed by "n" refer to notes.